CULTURAL INDUSTRIES.ca

CULTURAL INDUSTRIES.ca

Making Sense of Canadian Media in the Digital Age

Edited by

Ira Wagman and Peter Urquhart

James Lorimer & Company Ltd., Publishers
Toronto

James Lorimer & Company Ltd., Publishers acknowledges the support of the Ontario Arts Council. We acknowledge the financial support of the Government of Canada through the Canada Book Fund for our publishing activities. We acknowledge the support of the Canada Council for the Arts which last year invested $24.3 million in writing and publishing throughout Canada. We acknowledge the Government of Ontario through the Ontario Media Development Corporation's Ontario Book Initiative.

 ONTARIO ARTS COUNCIL
CONSEIL DES ARTS DE L'ONTARIO

 Canada Council   Conseil des Arts
for the Arts   du Canada

Canada

Cover design: Tyler Cleroux
Cover images: iStock

Library and Archives Canada Cataloguing in Publication

Cultural industries.ca : making sense of Canadian media in the digital age / edited by Ira Wagman and Peter Urquhart.

Includes bibliographical references and index.
Also issued in electronic format.
ISBN 978-1-4594-0273-7

1. Cultural industries--Canada. 2. Canada--Cultural policy.
I. Wagman, Ira II. Urquhart, Peter, 1967-

HD9999.C9473C3 2012   338.4770071   C2012-903777-X

James Lorimer & Company Ltd., Publishers
317 Adelaide Street West, Suite 1002
Toronto, ON, Canada
M5V 1P9
www.lorimer.ca

Printed and bound in Canada.

Contents

Foreword

Michael Dorland

For three decades now, Canadian scholars have grappled with the state of the cultural industries in this country, particularly their English Canadian components. What goes on with the Quebec equivalents still remains cloaked in its *différences*.

In 1983, Lorimer published Paul Audley's *Canada's Cultural Industries*. Single authored, it was the first book to anchor its analyses in Statistics Canada data, albeit with a five-year time-lag, but anchoring various unfocused worries about the state of the cultural industries in quantitative form. Operating within a cultural nationalist paradigm, Audley mapped out some of the key aspects of the reality of Canadian cultural production: for example, in film production, a plethora of small, short-lived firms that produced a film or two in a distribution and exhibition market overwhelmingly controlled by large American corporations.

In *The Cultural Industries in Canada* (1996), twelve contributors and I attempted to empirically track the changes that had occurred since the publication of Audley's book, gesturing toward elements of a policy and production apparatus that I characterized somewhat wishfully perhaps as "post-nationalist." This was a way of flagging forms of integration of Canadian cultural production within strategies that were no longer only Canadian but becoming increasingly "globalized."

As Wagman and Urquhart observe, each book about the cultural industries has been concerned with some form of cultural anxiety: Audley with the threat to cultural sovereignty posed by the overwhelming capacities of American production; my contributors and I with the role of the state policy apparatus in encouraging forms of globalizing integration; and

Wagman and Urquhart with the fear that studying profit-driven enterprises could have deleterious effects upon non-commercial or cultural bonds.

In other words, each book tries to do two things simultaneously: to identify changes on the ground since the previous edition in the genres of cultural production, here especially digitization and new media, but also to provide new directions for *actual* cultural industries research. How does one *do* policy research or research into companies in the absence of adequate industrial data?

The methodological turn offered by *Cultural Industries.ca* is one of its great strengths and a credit both to the editors and to the contributors. It is as well an important illustration of what scholarly research can bring to the study of its objects. In so doing, it challenges other scholars to further add and refine.

A Chinese proverb says, "Don't wait for Heaven to throw you a meat pie." This new examination of our cultural industries not only has ceased to wait for Heaven, it offers its readers plenty of meat to chew over.

Introduction

Ira Wagman and Peter Urquhart

In this book, experts from across Canada offer their reflections on the current state of our cultural industries, those sectors of the Canadian economy devoted to the production, distribution, and exhibition of various forms of popular culture, entertainment, news, and information. The convenient name we once used for these industries—the mass media—is less viable in the Internet age where audiences have become fragmented, companies that used to simply provide access to media content now deliver it themselves, and many new stakeholders have become part of the landscape. In fact, one could argue that "mass media" was never the right term in the Canadian context. Most domestic media outlets have constantly had to compete with firms based in the United States or elsewhere to attract audiences, and the idea of attracting a "mass audience" for Canadian cultural works has almost always been a pipe dream. Consequently, we use the term "cultural industries" as a convenient container to fit a wide range of stakeholders, such as video game producers, film distributors, recording studios, bookstores, and cellular phone services, among others.

One of the distinctive features of this book is a focus on methodology. We are not only interested in reporting on recent developments and challenges in Canada's cultural industries but also in questioning how to study the cultural industries in Canada. This book will have particular appeal to those who study these industries as well as to students who would like to study them but are unsure of how to begin. For this first group, the collection represents a call for a broader discussion about cultural industries methodology; for the second, it serves as a jumping-off point for undertaking research in this area of study.

There is an urgent need for an up-to-date evaluation of our cultural industries. Since the publication of Michael Dorland's *The Cultural Industries in Canada* in 1996, they have undergone significant changes. Some of these changes include the emergence of the Internet and the proliferation of new devices, from laptops to tablets, that allow for the distribution and exhibition of content. They also include the expansion of mobile communication through cell phones and smartphones that act as miniature computers, screens, and sound systems. The digitization of media forms once seen as distinct has created conditions for convergence. Content once intended for television, for example, can now be accessed on a variety of different platforms—laptops, tablets, smartphones—and the same is true for other media. Meanwhile, the proliferation of broadband networks and high-speed Internet transmissions has aided in the circulation of media texts on a global scale. Finally, Canada's transition from a manufacturing economy to a "knowledge economy" has put greater emphasis on the importance of intellectual property as a key structural element within the cultural industries. One might argue that media industries are always in transition, changing and adapting to new business practices, technologies, and policy developments. However, it is important to point out the extent to which digitization, globalization, and fragmentation of the cultural industries worldwide are having a transformative effect on the way creative work is produced, distributed, and consumed.

What is interesting about these developments is how they are playing out in the Canadian context. Like others, we note the tremendous tensions not only between new and old media forms in Canada, but also between new and old media stakeholders. Rather than one replacing the other, new and old players stand astride each other in the environment of abundant media content we now live in. While Canadians now get much of their content from companies like iTunes or through online activities that fall outside the rubric of cultural policy, such as the video-streaming service Netflix, there remains a strong and viable cultural policy apparatus that supports, to varying degrees, the production, distribution, and exhibition of domestic cultural content. While many predicted that the arrival of the digital age spelled the end for organizations like Canada's broadcasting regulator, the CRTC, the regulator remains an important institution in determining the shape and direction of Canada's airwaves. Newspapers and television networks still command large audiences even if their numbers have diminished somewhat. While the shrinking or frozen budgets of public institutions like the CBC and NFB continue to garner a lot of critical scrutiny, those two institutions have been among the most innovative in the areas of new media, providing popular and dynamic platforms to distribute their music, movies, and new interactive media. Although the Internet was supposed to serve as a challenge to Canada's established media institutions, this challenge has not materialized. Instead, we have witnessed a consolidation of media ownership in ways never before seen in history, with companies owning media properties across different platforms.

With these conditions serving as the industrial backdrop to this book, the question we consider is how we can go about studying the cultural industries in this context. We have observed a methodological gap between *research on* the cultural industries and *research on how* one should do research on the cultural industries. In a recent collection, Havens and Lotz (2011) introduce the concept of the "industrialization of culture" model, in which a media text—a television show, for example—is formed by a series of factors including social tastes, mandates, conditions, practices, texts, and ideas from the public arena. Others, like David Hesmondhalgh (2007), offer an approach to the study of cultural industries that is based on what he calls "an eclectic methodology," mixing aspects drawn from cultural studies, sociology, communication studies, and political economic approaches to media. Both approaches are useful in providing a broader understanding of how to make sense of the study of the cultural industries. We think such approaches need to be complemented by more modest reflections on specific kinds of research methods. If a researcher wanted to perform what we call "policy analysis," where would this person start? How can we account for a key idea in the study of culture, that of media ownership, if we don't know how to find that information?

Research into Canada's cultural industries has been hampered by a paucity of industrial data. This is, in part, due to the fact that the primary collectors and distributors of those data have been various institutions of the federal government. It is through Statistics Canada reports, or reports of the CRTC, or through studies from Royal Commissions, or Senate reports that much of the data on the cultural industries become collected and made available. However, there now exist a number of different data sources that collect a range of different information about how Canadians consume various kinds of cultural works. The industry publication *Playback* releases weekly reports on top-selling DVDs, or box-office figures, or ratings of Canadian television programs. Websites like comScore or Alexa offer, usually with a grain of salt, information on popular websites on a country-by-country basis. Cynopsis.com, an aggregator of media industry news, regularly collates data from different sources into articles on a wide range of cultural industries.

Data do not account for everything, though. Research on the cultural industries is a highly interdisciplinary endeavour, drawing on policy reports, annual financial statements, research from private firms, newspaper accounts, demographic information, marketing theory, and so on. In the current context, one needs to be aware of, for example, issues having to do with copyright, the politics of search engines, and the power of social networking. The current emphasis on the challenge of research in the age of "big data" may be a valid concern, but it only draws attention to the fact that this kind of research has always been "big" and needed to be drawn from a variety of different sources.

We also believe that an issue facing the study of the cultural industries has been attitudinal. So much of the thinking about the cultural industries is tied up with three very different strands of thinking. The first represents anxieties associated with modernity and, more importantly, with the role that media play in representing, assisting, or replacing humans in the functions we once performed without them. The second represent anxieties associated with any industrial sector: that it ebbs and flows, that some firms are healthier than others, that various kinds of external pressures influence its future viability, or that new entrants threaten to challenge its hold on a given sector. The third are anxieties associated with nationalism. Here, the fear is that Canadian values will somehow disappear or be taken away from us due to the effects of foreign media. We can see in each of these cases an idea associated with theft, or to paraphrase an idea from Kevin Dowler (2011), the inherent belief that media steal things from us (6).

These conceptions have had a powerful role in framing the way we see the cultural industries. The problem, however, is that none of these concerns begin with the recognition that the providers of Canadian software, television, radio, and so on are commercial operations whose raison d'être is to generate profit. This commercial orientation may be moderated somewhat by the fact that there exists an elaborate policy apparatus in place to ensure that their objectives be offset by investments in Canadian cultural production, or commitments to air Canadian content, or limits on their ownership of media properties. That said, though, those in the cultural industries perceive that they are in a competitive marketplace, and view their own situation through the prism of global developments. Again, how this situation is then mediated through localized issues, such as policy frameworks, legal restrictions, ownership structures, and audience perceptions, is something the contributors to this collection explore in considerable detail.

Part One of this book questions the features of the new landscape that Canada's cultural industries must now operate in. We can draw attention to them by pointing to discussions by some of the contributors to this book. In Peter Urquhart's chapter on film and television, he argues that the lens through which one examines the relative success or failure of these industries directly affects the conclusions drawn, and proposes that this sector has actually been far more successful than is typically claimed. The interrelationship between radio and sound recording is the overarching theme in Richard Sutherland's article. In the two entries dealing with print media, Jeff Boggs's article on book publishing and Chris Dornan's chapter on newspapers and magazines, the changing nature of the audience and technologies serves as the focus. Greig de Peuter's chapter draws attention not only to a new and important area of the cultural industries, that of video game production, but also to new kinds of working conditions facing those who are employed in "the creative economy." Finally, Daniel Paré reminds us of the fact that all the content discussed in the previous chapters must now be distributed through an elaborate telecommunications system, and how much old companies and old ideas about media form a vital part of the distribution of new media content.

Part Two of this collection turns our attention to questions of methodologies. The first three chapters in this section are devoted to making sense of how to research the cultural industries. Zoë Druick reminds readers that what we call "the cultural industries" has changed over time. How we define something determines not only what we want to study, an epistemological question, but how we want to study it and the approaches we use to do so. In the next chapter, Dwayne Winseck takes a personal view, showing the complex and contradictory forms of knowledge used to make sense of one of the more complicated issues in political economic studies of the media, that of ownership. Finally, Jeremy Shtern argues that in Canada many people practise "policy analysis," but few have ever thought of this form of analysis as a methodological approach, one driven by certain choices, ethical issues, and resources.

The final three chapters use historical methods to draw attention to some forgotten histories of Canadian cultural industrial activity. These chapters do not call for historical reflection; rather, they posit what we believe are three different questions relevant to the study of Canada's cultural industries. Olivier Côté draws attention to English–French tensions in cultural production in Canada and demonstrates the insights provided by using a "production studies" approach. Mark Hayward's institutional history of multicultural television draws attention to the largely forgotten history of multicultural media in this country and to the tendency to study the cultural industries only in terms of the main players. Finally, in Sandra Gabriele and Paul Moore's chapter, the question is about how old and new media interact. While their example is drawn from the arrival of radio into the newspaper business, it remains germane as an approach for us to study as new technologies—from tablets to cell phones—change, or at least challenge, the ways in which Canada's cultural industries operate, just as radio challenged newspapers in the 1920s.

As the cultural industries now operate in an environment of abundance, we recognize that we cannot capture all of their activity in this collection. Many areas remain unexamined here. We also note the continued underdevelopment of audience research, both as an object of study and as a research methodology. That said, we believe that with the information and methodological observations offered in this book, readers will be in a much better position to undertake that research as well as to pursue numerous other areas of inquiry in an attempt to map Canada's cultural industries at a time of considerable change.

References

Dowler, Kevin. 2011. "Why Study Communication?" In *Intersections of Media and Communication*, edited by Will Straw, Sandra Gabriele, and Ira Wagman, 3–20. Toronto: Emond Montgomery Publications.

Havens, Timothy, and Amanda Lotz. 2011. *Understanding Media Industries*. Oxford: Oxford University Press.

Hesmondhalgh, David. 2007. *The Cultural Industries*. London: Sage.

PART ONE
CULTURAL INDUSTRIES IN TRANSITION

1

Film and Television: A Success?

Peter Urquhart

Success

Despite a decades-long rhetoric of failure surrounding the Canadian film and television industries and, despite the potentially devastating effects of digital transformations (particularly via new distribution platforms) presumed in some quarters to be looming, these sectors of the Canadian culture industries continue to flourish. With the occasional minor hiccup, for example, related to measurable exterior factors such as the global recession of 2008–2010, these cultural industries continue to perform very well in both industrial and cultural terms. This chapter will examine the structure of the industries, consider the relationships between film and television producers, disseminators, government policy and audiences, and frame these considerations in a larger argument about the *perception* of failure which has surrounded these cultural industries in the national imaginary for the past forty years or more despite their very strong performance.[1]

First, allow me to demonstrate the phenomenal success of the film and television industries. According to Statistics Canada figures released in the spring of 2012, the sector they call the film, television, and video industry enjoyed operating revenues of $3.3 billion in 2010, a 7.6 per cent increase from the previous year. Television production led the way, accounting for 53.9 per cent of these revenues, followed by feature film production at 15.9 per cent, commercials at 13.3 per cent, and the industrial/corporate video sector came in at a surprisingly high 7.2 per cent. On the expenses

---

1 A good example of the miserablist position on this industry is provided by Manjunath Pendakur's *Canadian Dreams and American Control: The Political Economy of the Canadian Film Industry* (1990).

side, the same StatsCan report shows $686.1 million was paid to Canadian workers in salaries, wages, and benefits in the production industry alone in 2010. While profit margins were relatively meagre in the production industry at just under 2 per cent, the distribution side enjoyed a nearly 25 per cent profit margin in 2010 on approximately $2 billion in revenue versus nearly $1.5 billion in operating expenses (with these figures excluding broadcasters as "distributors" of content). While the vast majority of Canadian distributors accrued their revenues from the distribution of non-Canadian product in the domestic market, nearly $180 million of these revenues resulted from domestic distribution of domestic products (Statistics Canada 2012).

Furthermore, according to a 2011 study by the Canadian Media Producers Association (formerly the Canadian Film and Television Producers Association), the film and television *production* industry employed 128,000 full-time equivalent workers (FTEs), 53,000 of those directly in production with the remainder in related spinoff employment. Production alone contributed $3.07 billion to the Canadian economy, with a further $4.39 billion generated by spinoff activity (CMPA 2011). Remember, these are figures for the production aspect of the film and television industries alone and do not include revenues from exhibition, broadcasts, and so on.

The data I have just provided should hopefully convince the reader that, at least in economic terms, the Canadian film and television industries are very successful. Shortly, on the one hand, I will discuss these two sectors in isolation from each other, and on the other, demonstrate the logic of their interdependencies. Finally, I shall provide arguments and evidence for the relative cultural success of film and television in Canada as well.

Private Television

Three assumptions surround everyday discussions of the contemporary private television landscape in Canada. The first is that while Canadians produce and sell a great deal of content both domestically and internationally, Canadian audiences overwhelmingly prefer American content, as audience measurement seems to clearly demonstrate. My students call this the "Canadian TV sucks" problem. The second assumption concerning the industry is that concentration of ownership in the Canadian media industries at large, but particularly in the television broadcasting and transmission sectors, has accelerated in the past few years, leaving almost every single one of the nation's television channels in the hands of only four companies—Shaw, Quebecor, Bell, and Rogers—and that this concentration of ownership has consequences for the type of content produced and broadcast over Canadian airwaves. We might call this the "oligopoly problem." Finally, the third assumption about television in Canada today is that the rapid proliferation of alternative platforms for the reception of television content, via services like iTunes, Netflix, online providers both pirate and legitimate, and, especially, so-called third-screen services, which provide television content to mobile phones and tablets, will dramatically alter not only the revenue models for conventional broadcasters, but also the entire structure of the television production and distribution industry in Canada. We might characterize this concern as the "(cable) cord-cutting freak-out." This section shall take up each of these three assumptions, in order.

But first, a few broadcaster-specific statistics are in order: here, we find even more spectacular economic results than I cited above for industry-wide production. Despite the cord-cutting freak-out, after the first down-turn in broadcasting profits in thirty years in 2009, caused by the dramatic decrease in advertising revenue resulting from the global recession, 2010 saw Canadian broadcasters enjoy an 8.1 per cent increase in revenue to $7.1 billion. Conventional broadcasters saw profits improve slightly while specialty and pay services continued their very high rate of profitability. In fact, the 2010 figures demonstrate a continuing trend over the past several years of specialty channels consistently outperforming conventional channels in terms of profitability, and now they are approaching conventional channels in terms of their overall percentage of broadcasting revenue. If this trend continues, and there is no reason to expect that it will not, specialty and pay services will very soon surpass conventional broadcasters in their percentage of revenue generation in the broadcasting industry. Advertising, unsurprisingly, continues to contribute the most revenue to broadcasters' bottom line (just under half), but with the continuing rise of specialty services, subscriber fees have increased from 22.5 per cent of total broadcasting revenue in 2000 to 31.7 per cent of the total in 2010 (Statistics Canada 2011).

So, if television production created billions of dollars in economic activity and if the broadcasters generated over $7 billion on top that, where does the "Canadian TV sucks" problem come from? Teaching Canadian television to students in Quebec, Ontario, and the UK over the past dozen years, I have regularly encountered the assumption from them that it is the "low budgets" of Canadian television, certainly as compared to American TV, that make our production pale in comparison. But if billions are being generated by this industry, and in a relatively small country (i.e., in a relatively small market), is it the case that Canadian television is "cheap" and therefore, at least to students—and perhaps to the everyday audience as well—less good? In a later section of this chapter, I am going to make an argument for the cultural "success" of Canadian television, but for now, let's consider some more data. According to the Bureau of Broadcast Measurement, in 2011, seven of the ten most-watched broadcasts in Canada were Canadian productions, with only the Super Bowl (third place), the Academy Awards Ceremony (fourth place), and Ashton Kutcher's debut in the newly revamped tigerblood-less *Two and a Half Men* (tenth place) originating in the US. Some of you doubt this, perhaps. You wouldn't if I told you that every one of the seven Canadian entries that were among the top ten most-watched broadcasts of the year was a hockey game. Six of the seven games of the 2011 Boston–Vancouver Stanley Cup finals ranked in the top ten for the CBC, as did the finals of the World Junior Hockey Championships for TSN (Digitalhome.ca 2012).

You will notice that nine of the top ten most-watched TV broadcasts in Canada in 2011 were specials, with only one conventional series program, *Two and a Half Men*, cracking the list. The picture for Canadian TV looks rather different than the above example when we scrutinize the top ten most-watched series programs in the country. All ten of these are American network shows, such as the most watched, *The Big Bang Theory*, and the second most watched, *American Idol*, followed by *Survivor* and *House*. In fact, the only Canadian program that was not a hockey game or a newscast to consistently appear in the weekly top thirty most-watched programs was the CBC's *Dragons' Den*, which though a regular series is decidedly not a comedy or drama of the sort which has typically characterized the most popular American TV series (until the reality television breakthrough of the recent past). *This* is the problem my students are talking about when they declare that Canadian TV sucks. They mean Canadian *comedies and dramas* on TV suck. While there have been some modestly popular Canadian-produced dramas and comedies over the past few years—CTV's *Flashpoint* and Global's *Rookie Blue* are good examples—it remains the case that despite its volume of economic activity, the Canadian television system seems unsuited to or incapable of competing with American networks in this particular arena.

I'd like to take up a point here raised in this book's introduction, that of "attitudinal" responses to Canadian cultural industries, especially as attitudes toward Canadian television are concerned. In that the industry is demonstrably successful from an economic standpoint, and in that a high percentage of the most-watched broadcasts of 2011 were Canadian in origin, and in that hundreds of millions of dollars go into the creation and dissemination of Canadian comedy and drama programming (providing space on the dial for the local option), why are so many Canadians, and scholars and critics as well, distressed by the fact that Canadian drama and comedy are less popular than American fare? I suggest this is an attitudinal problem more than it reflects any particular failure of an otherwise extremely successful industry. The success that American networks have had selling their content internationally has to do with economies of scale—chiefly the size of their own domestic market—which provides profit for successful American shows in the US alone, therefore allowing the producers to sell these shows in other territories at massively discounted prices. Thus, it costs CTV far less to buy the rights to broadcast *The Big Bang Theory* in Canada than it would to produce its own programming of a comparable degree of gloss and polish (and, some might say, a comparable degree of accomplishment). In addition to this, these discounted American shows have the added benefit of built-in promotional mechanisms via the American entertainment industry self-promotion machine that operates widely in Canada as well: infotainment on television, magazines, print campaigns, celebrity interviews, and so on. In the face of all this, we should be neither surprised nor distressed that Canadians choose American comedy and dramatic television programs overwhelmingly over Canadian versions. We have regularly available, sometimes popular, sometimes not, Canadian programming to choose from, which is almost always much more modest in scale and sometimes still finds an audience. The only reason this situation is regarded as a failure of the system is attitudinal—that in this one particular aspect of the entire televisual system, Canadian television seems inadequate, despite a deck stacked egregiously against us.

I turn now to the second assumption about what is wrong with the Canadian television industry today—the oligopoly problem. Let me first spell out the perceived difficulty. For an audience to experience a program, there are three separate but related industrial elements: production, distribution, and transmission. The latter two tend to dictate conditions to the first; in Canada in 2012, distribution and transmission are almost entirely controlled by Shaw, Quebecor, Rogers, and Bell. Together these companies, through their various broadcast arms, own nearly every channel on the dial, both conventional and specialty, *and* they entirely dominate the transmission of TV across the country through their cable and satellite divisions. For example, the distributors that Shaw owns—that is, the networks and specialty channels—include the nationwide Global network,

Showcase, HGTV, History Television, Food Network, and more than two dozen others. *And* Shaw is the second leading cable company in the country with approximately 30 per cent of the market share, transmitting all the channels it owns in addition to dozens of others, many of which are owned by the companies it competes with in the cable and satellite business. This intensely oligopolous relationship—again, nearly every broadcast entity is owned by, and nearly all cable/satellite customers are subscribed to, one of four companies—results in some very peculiar relationships in Canada's highly regulated broadcasting environment. For example, one condition set by the CRTC which Canada's cable companies must adhere to is that a percentage of gross annual revenues be returned to the Canadian Media Fund (CMF), which distributes this money to broadcasters to commission or purchase Canadian television content, in exchange for a highly regulated cable environment, which basically assures these firms, and the few others which exist, a guaranteed customer base. Both Pierre Karl Péladeau, the CEO of Quebecor, and J. R. Shaw, the (now former) CEO of Shaw Media, have complained loudly in the past about this provision, which forces them to hand some of their cable revenues over to the CMF to distribute to broadcasters based on its own criteria (Bailey 2010). However, since all four of the main cable/satellite firms are *also* broadcasters, much of this money is returned to the broadcasting arm of the same company through the CMF to spend on original Canadian programming. In other words, for example, Shaw the cable company hands over several million dollars to the CMF, which distributes it to broadcasters, including Shaw-owned channels. In the incredibly protected cable and broadcast market, these four firms are assured large profits at the same time that they control all aspects of the industry: they have a huge percentage of the cable and satellite subscribers and own most of the channels and, therefore, can dictate production as well since they invest a largest percentage of cash in commissioning new production. Furthermore, the power of the oligopoly extends into the realm of new media, and because these companies own the online rights domestically for the content they distribute via broadcast, Canadians are barred from services like Hulu and Comedy Central and are forced to use *their* personal apps to get certain types of third-screen programming.

One group of stakeholders that objects to this state of affairs is documentary producers. The volume of documentary production has steadily decreased over the past few years, and one explanation for this is that broadcasters are commissioning less of it to be produced—television being the primary exhibition venue for documentaries. In the latest iteration of their studies of the documentary industry in Canada, the industry lobby group DOC (Documentary Organization of Canada 2011) explicitly cited the concentration of ownership and the tightly oligopolous situation in television broadcasting and transmission as one of the causes for the decline. As documentary is frequently seen to be serving the public interest in ways that comedy and drama are not, DOC is concerned with its survival and argues that the situation I have just described needs to change if documentary production in Canada is to survive.

Now we come to the "cord-cutting freak-out," the last of the three assumptions about television in Canada today. In light of the tightly regulated, state-sanctioned oligopoly which describes the industrial model of television in Canada, it seems reasonable to assume that television is endangered by the declining advertising revenue that should naturally result from a large migration away from traditional television watching toward audiences acquiring and watching television content on other, non-advertising-supported platforms such as Netflix, on-demand services, and via "third screens," including mobile phones and tablets. As well, illegal downloads of television content via BitTorrent and a wide array of websites sharing pirated content is at least as much of a bogeyman for the industry as were Napster and other MP3-sharing sites for the music industry.

Again, some data might be useful here. First, according to a 2011 report in *Playback*, the Canadian film and television industry publication, only 0.7 per cent of Canadians cut the (cable) cord by cancelling cable TV service and moving to some combination of over-the-air reception (which is, since the switch to the digital signal, now a much more desirable option than was over-the-air analog)—Netflix, Apple TV, and so on. On the other hand, new cable subscriptions in 2011 did fall by half in 2010—from about 220,000 new subscribers to 110,000 (Vlessing 2012a). At least in the eyes of the broadcasting industry and the regulator (CRTC), Netflix is the largest cause for concern. The US-based film and television streaming service has over one million subscribers in Canada as of spring 2012, but, and this is the significant detail for our purposes, the *Hollywood Reporter* tells us that a report from The Media Technology Monitor shows that more than four out every five Canadian Netflix subscribers also subscribe to a cable television service, with the vast majority of these customers having no plans to cut their cable cords anytime soon. It would appear that, like so many claims of a "digital revolution," the cord-cutting freak-out is overblown at best in the tightly regulated, highly protected oligopolous environment that is the Canadian television industry. At least for the time being.

Public Television

If not for the interesting moves being made by British Columbia's Knowledge Network and the existence of a large provincial public broadcaster such as TVO, this section could have simply been called "The CBC." Knowledge Network's move in 2011 to acquire the BBC Kids channel is of interest because Knowledge Network explained it as a "natural" expansion of its interests since children's programming is central to Knowledge's brand (much as it is for TVO), while, on the other hand, it is an unprecedented move for a public broadcaster to expand into the realm of private television (Knowledge.ca 2011). Knowledge has also undertaken to form partnerships with other provincial pubcasters, including TVO, to collaborate on commissions and to share resources in other ways.

While CBC television remains centrally important to this sector of Canada's cultural industries, its importance might be waning. With the massive expansion of options on the dial and with private conventional and specialty broadcasters having become so successful at capturing audiences, other than hockey games, and the occasional moderately popular realty show such as *The Battle of the Blades* or *Dragons' Den*, the CBC's struggle for viewers gets tougher all the time. Which is not to overlook its important position not only in the national imaginary, but also in providing television service to parts of the country that are undesirable to private broadcasters because they are unprofitable, and, one could argue, also in providing regionally specific content to the entire country. Thus, in the 2011–2012 television season we saw such fare as *Heartland*, set in Alberta; *Republic of Doyle*, set in Newfoundland; *Little Mosque on the Prairie*, set in Saskatchewan; and *Mr. Dee*, set in Toronto. All of these shows enjoy or have enjoyed modest popularity, each achieving at least one million viewers on occasion, a benchmark of popular success in Canadian television. In the past, we've seen other popular CBC shows set unambiguously in other places, too: *Da Vinci's Inquest* made much of its Vancouver setting, for instance. These nationwide broadcasts of regionally specific content can be seen as part of a concerted attempt on the part of the CBC to fulfill the part of its mandate to "reflect Canada and its regions to national and regional audiences" (CBC/Radio-Canada 2012). It is perhaps because of the availability of such fare, and the fact that CBC television airs a much higher percentage of Canadian content in prime-time than do the private networks, that the CBC maintains its position in the national imaginary, even if its audiences are relatively small. According to the CBC's own research (which must be taken with a grain of salt, admittedly), 92 per cent of Canadians regard the CBC as an "essential service."

Film in Canada

In 2010, producer Don Carmody broke his own long-standing record for the highest-grossing Canadian film of all time when the fourth install-ment of his videogame-derived franchise *Resident Evil: Afterlife* surpassed *Porky's* (a film Carmody produced way back in 1981) in total box-office take, eventually going on to gross around $300 million worldwide. This film and its huge international success serve as a very useful example when we begin to discuss the contemporary Canadian film industry, which, as I showed at the beginning of this chapter, along with television production, is very successful from an economic standpoint. This film is representa-tive at the present moment in that it demonstrates, on the one hand, the commercially successful industry, and, on the other, a variety of anxieties and misgivings on the part of many in Canada—those in the industry themselves and those not—about the *kinds* of films being made in Canada. Most of these anxieties and misgivings can be distilled into fretting over two categories of film production: international co-productions and what we call the "service" production industry (productions in Canada by for-eign-owned firms, often known as "runaway" productions). Both of these categories of production tend in the main to be films that unsettle some people because, it is argued, either their "Canadianness" is totally absent, in the case of runaway productions, or it is often difficult to discern, in the case of international co-productions. Let's look at these categories of production one at a time.

According to a report by the CMPA, based on data collected from the Canadian Audio-Visual Certification Office (CAVCO, administered by the Department of Canadian Heritage), among other sources for some data sets, domestic feature film productions in 2010–2011 contributed $306 million of the approximately $5.49 billion in total film–television production revenue in Canada. So while this $306 million was spent on making Canadian films, the total volume of service production in the same period was $933 million, up over 35 per cent from the previous year. This volume of production activity also generates employment, with approximately 7,100 jobs in Canada resulting from domestic film production and three times as many from service production. Among the prominent and successful domestic features produced in 2010–2011 were *Incendies*, Denis Villeneuve's Academy Award-nominated thriller, and the gigantically successful *Resident Evil: Afterlife*. In the same period, promin-ent service productions (again, these are "runaway productions," almost always Hollywood films shot in Canada using Canadian crews) included *Contagion* and *Planet of the Apes*.

There has been a long-standing concern about this state of affairs, a worry that it is unhealthy that service production for Hollywood studios represents about three times as much production activity in Canada than does domestic production. Often the metaphor of the "branch-plant" is invoked here, where Canadians are reduced to minor roles in the creation of Hollywood films, thus inhibiting domestic production, since our crews and locations are taken up with doing Hollywood's bidding, for union scale. I would like to examine some of the premises that underlie the fear and disdain surrounding service production. The first of these has to do with labour itself. With so many skilled technicians and craftspeople—from set decorators to gaffers and grips to drivers to caterers—taken up with Hollywood shoots in Toronto or Montreal or Vancouver (where the vast majority of service production takes place), one argument goes that these skilled Canadians are therefore unavailable to work on domestic productions. However, because of the project-based nature of film production—a thirty-day shoot means thirty days of employment for a set dresser, say—I suggest that a thriving service sector is actually a prerequisite for the existence of a healthy domestic feature film industry and not an impediment to one. This is because a large and highly skilled workforce has been created through the expansion of the service sector, and this professional workforce is available for domestic production as well. Because of the much smaller domestic industry, without service production it never would have generated such a large pool of skilled film production specialists. Furthermore, some of the other results of the burgeoning service sector are other infrastructural necessities for the existence of a vibrant domestic industry, including things such as large professional soundstages and advanced post-production houses, both of which Canada has, and again because of service production, which the domestic industry can now make use of. Service production creates employment, generates capital, and contributes greatly to the production infrastructure—all in the aid of the domestic industry. It should therefore be seen as a vital part of the Canadian film production industry, even if the films created by this sector are not Canadian films.

As a final note on the strength of the service sector, let me correct one other misconception. It has been assumed for years in the literature on runaway production and even in the national press that a major explanation for the surge in service production in Canada since the 1980s was because of the currency differential: that the low Canadian dollar made production here very desirable to American firms. While it may be the case that a 65-cent dollar *was* attractive to American companies, we now know that this was at best an insignificant part of the reason Hollywood was making films in Canada. The past few years have seen currency parity with service production massively increasing, not decreasing as one would have expected if the low dollar explained runaway production. In fact, the CMPA report cited earlier contains a graph that shows service production consistently increasing since the 1980s with no correlation whatsoever to the value of the Canadian dollar. A more likely explanation, therefore, for the success of the Canadian service sector is that we have great crews, great locations, and great talent for making movies and television shows.

The other aspect of film production in Canada that raises the eyebrows of cultural nationalists is international co-production. Canada has treaties with over fifty countries: these agreements allow producers in two or more nations to pool their creative and economic resources and to take full advantage of industrial incentives provided by their own national (and provincial or other) governments as if the co-production were a national production. So, for example, the David Cronenberg–directed *Cosmopolis*, with its international cast that includes Juliette Binoche, Robert Pattinson, and Paul Giamatti, is an official treaty co-production between Prospero Pictures (Canada) and Alfama Films (France) which is treated as a Canadian film for the purposes of government incentives in Canada and as a French film by the French government for the same purposes. One can see why treaty co-production might be seen as a problem for proponents of a "culturally" national cinema. In fact, European commentators even coined a new term for multi-nation European co-productions, which were seen to homogenize and even eradicate the national specificities of the European nations involved in the co-pros: they were given the unappealing label "Euro-pudding." Likewise in Canada, some fear that international co-production treaties—because they are so popular with Canadian producers, providing, as they do, opportunities for much larger budgets than would otherwise be possible, and for creative collaborations—water down the national specificity of our cinema and work against the creation of a truly national cinema.

There are sound rebuttals to these worries. The first is that co-production, though a very popular financing tool, is not a requirement, it is only an option. Producers able to raise sufficient capital from Canadian investors do not need to seek out international partners to get their projects made, but those who do wish to are in a very advantageous position. Secondly, feature films as cultural artifacts are very frequently profoundly international texts to begin with—observe the international movements of the key creatives, stars, directors, cinematographers, and so on and of border-crossing plots and themes, for instance. Co-production, therefore, can be seen not only as a valuable tool to producers, but also as recognition of the already-existing conditions and relationships. Finally, international co-production partners provide much easier access to foreign markets for Canadian films, thus increasing their profile internationally and, very likely, their audience.

Other Measures of Success in the Canadian Feature Film Industry

In light of the frequency with which the apparent failures of the Canadian feature film industry are pointed out—usually via the measurement of "screen time," the data that show year after year how Canadian films garner about 2 or 3 per cent of screen time in Canadian cinemas—I recognize that I risk appearing Pollyannaish by pointing to the various measures of success in this cultural sector. In response, let me refer once again to the question of "attitudinal" responses to the Canadian cultural industries at large and to feature film production in specific. As Charles Acland (1997) has demonstrated, the primary discourse that has surrounded the Canadian feature film industry since it emerged in the 1960s has been of failure and absence (281). Canadian features make it onto our screens pathetically seldom and audiences for them are woefully absent, or so goes the dominant view. What if the "attitude" toward measures of success and failure shifted somewhat? Much like the attitudinal problem of responses to Canadian television's relationship to American television, what if we acknowledged that very few minor national cinemas compete in the marketplace in any meaningful way with Hollywood and that the problem of proximity, language (and so many other markers of deeply shared culture), and a small domestic market compound the difficulties for Canadian cinema? What if we shift the goalposts somewhat and adopt a more nuanced yardstick for the success of Canadian cinema than "screen time" or mere box office performance?

Interestingly, this is exactly what Telefilm Canada, the federal government agency that is the primary source of government support for the feature film industry, has recently done. The centrepiece of the agency's 2011–2014 corporate plan, *Fostering Cultural Success*, is a new "Success Index" for the measurement of the efficacy of Telefilm's participation in the Canadian film industry (Telefilm Canada 2012). Until now, the *only* measurement of success was based on box-office figures. These could be easily counted, but it is increasingly clear that for a number of reasons simply counting box-office receipts is a very unsophisticated, and an unsatisfactory, data set. First of all, it has been the case for a long time that Canadian films derive significant revenue, and often a majority, from sales to broadcasters. Far more people see Canadian films on television than they do in cinemas. Secondly, "ancillary" distribution through things like DVD sales/rentals and emerging digital distribution platforms such as Netflix and iTunes have become increasingly significant to a film's bottom line. In light of this, Telefilm's inclusion of these data in the first category of their new Success Index provides a much more accurate picture of a film's relative financial success than did the blunt instrument of simply counting the box-office take. Telefilm now counts this total commercial performance as 60 per cent of their Success Index. A further 30 per cent of the Success Index is devoted to what it calls "cultural" success: 15 per cent of this for selection at film festivals and 15 per cent for awards won in selected competitions, including the Oscars, the Cannes Palme d'Or, and so on. The final 10 per cent of the Success Index is given over to "industrial" success: measured by a ratio of public to private investment in the project. The higher the percentage of private participation, the higher the industrial score, under the assumption that the generation of private capital is not only industrially "healthy," but also demonstrates confidence in the market-worthiness of the project.

The adoption of this much more broad data set for measuring the relative success of feature films in Canada is useful for illustrating the attitudinal problem which has resulted in the consistent characterization of the Canadian film industry as a failed one. It has been the case for many years that Canada has produced films that have been seen (and perhaps even enjoyed) by hundreds of thousands more people than brute box-office figures suggested, chiefly on television, but also through other means. It has also been the case that Canadian films are regularly celebrated by the taste-makers of international festivals and awards. Canadian films are frequent recipients of Academy Award nominations (and even win some), and there is seldom a year where one or more are not successful at getting selected for prestigious festivals, including Cannes and Berlin. These things matter to the "success" of a film, even if its box-office performance is less than outstanding, because of the recognition they bring: recognition that Canadians make great films and that they make important contributions to culture at large and certainly to *Canadian* culture. Finally, the industrial measure is also a valuable addition to the discourse of success. Because of the very small domestic market, raising capital has always been a problem for Canadian film producers; knowing which films do generate private capital (and therefore entrepreneurial confidence in the likelihood of profit—or the likelihood that lots of Canadians might actually see the film) provides a level of nuance to the measurement which would be absent otherwise. For a hypothetical example: Is an arty auteur film which very few Canadians will ever see but which wins awards at the Cannes and Berlin film festivals more or less "successful" than *Resident Evil: Afterlife*, a film which was entirely privately financed and which was a worldwide hit? Clearly both films are "successful" to some degree, and a system of measurement that acknowledges this may go a long way to correcting the merely attitudinal, *habitually* attitudinal, problem that has created a cloud of "failure" over a cultural industry, which is actually extremely successful in many ways.

Finally, let us remind ourselves about indisputable signs of a vibrant, successful feature film industry. Economic success is evident from the data I have already supplied. Culturally and critically, there have been a hand- ful of demonstrably excellent films produced in Canada every year over the last three decades at least. In an article entitled "Genies 2012: Is This the Best Year Ever for Canadian Film?" *Toronto Star* critic Peter Howell quotes two informed observers, TIFF's Steve Gravestock and filmmaker Patricia Rozema, who both concur that the depth and range of the nominated films at the 2012 Genie Awards demonstrate an affirmative answer to the headline's question. The article describes the wide range of genres of the nominated films and quotes Gravestock on "how international this year's Best Picture field is, while at the same time retaining a distinct Canadian identity" (2012, E1). And finally, many of the Genie-nominated films found audiences in addition to critical acclaim, with *Monsieur Lazhar* and *Starbuck* having done particularly well at the domestic box-office. As Rozema observes, "Quebec is always healthy, but I suspect the rest of us are truly coming into our own as a movie-making/movie-loving nation. I predict we'll finally have a few English-Canadian commercial hits in the next couple of years. We are finally ready to like ourselves" (E1).

The Future?

The successes of the Canadian film and television industries today are largely reliant on the status quo industrial structure and policy appar- atus that has nurtured it. But, as we all know, the ground is shifting. The measurable successes of Canadian film and television cannot continue without adaptation to new conditions wrought by the ceaseless digitiza- tion of culture. In the face of this, it is notable that the main instruments of government involvement in the screen-based industries are beginning to adapt to this uncertain future. Since the Canadian Film Development

Corporation was reorganized into Telefilm Canada in 1983 (putting the television before the film), there has been no more significant analogous change than the reorientation in 2010 of the Canadian Television Fund (CTF) into the Canadian Media Fund (CMF). Whereas the previous body provided production incentives to the Canadian television industry, the CMF ties its support to two programs: the "experimental" stream and the "convergent" stream. The latter requires successful applicants to demonstrate a value-added aspect to production funding by distributing their project on at least two different platforms. One is almost always broadcast television, and the second must be what they call "rich, interactive content such as games, interactive web content, on-demand content, podcasts, webisodes, and mobisodes" (CMF 2012). This requirement for funding forces producers to deal with what looks to be the future of content delivery, even in a tightly regulated television market such as Canada's.

Perhaps even more interesting is the CMF's "experimental" stream that "encourages the development of innovative, interactive digital media content and software applications. Projects funded under the Experimental Program are to be developed for commercial potential by the Canadian media industry or public use by Canadians" (CMF 2012). At the same time, the venerable old National Film Board of Canada is devoting significant creative and economic resources to creating interactive digital media pieces and fewer resources to traditional forms of film. While NFB-produced work still gets nominated for Academy Awards, it now also wins Webbys for the NFB's foray into interactive work. Screen-based media are likely to change significantly in the future, and new interactive forms of storytelling, meaning-making, and art are emerging. Just as earlier policy instruments called into being successful film and television industries, current directions suggest that the policy apparatus is attempting to keep ahead of the curve.

References

Acland, Charles. 1997. "Popular Film in Canada: Revisiting the Absent Audience." In *A Passion for Identity: An Introduction to Canadian Studies,* edited by David Taras and Beverley Rasporich, 281–96. Toronto: ITP Nelson.

Bailey, Patricia. 2010. "Péladeau Threatens Court Action Against CMF." *Playback,* June 24.

CBC/Radio-Canada. 2012. "Mandate." Accessed April 27, 2012. http://cbc-radiocanada.ca/about/mandate.shtml.

CMF (Canadian Media Fund). 2012. Accessed May 29, 2012. http://www.cmf-fmc.ca/funding-programs/overview/.

CMPA (Canadian Media Producers Association). 2011. *Profile 2011: An Economic Report on the Screen-Based Production Industry in Canada.* Ottawa: CMPA.

Digitalhome.ca. 2012. "TV Top 10: What Canadians Watched in 2011." Accessed April 15, 2012. http://www.digitalhome.ca/2011/12/tv-top-10-what-canadians-watched-on-television-in-2011.

Documentary Organization of Canada. 2011. *Getting Real: An Economic Profile of the Canadian*

Documentary Industry, vol. 4. Toronto: DOC.

Howell, Peter. 2012. "Genies 2012: Is This the Best Year Ever For Canadian Film?" *Toronto Star*, March 12, E1.

Knowledge.ca. 2011. "CRTC Approves BBC Worldwide Canada and Knowledge Network Joint Venture for BBC Kids Channel" (March 29). http://www.knowledge.ca/press/bbc-kids-crtc.

Pendakur, Manjunath. 1990. *Canadian Dreams and American Control: The Political Economy of the Canadian Film Industry*. Detroit: Wayne State University Press.

Statistics Canada. 2012. "Film, Television and Video Production 2010" (March 22). http://www.statcan.gc.ca/pub/87-010-x/87-010-x2012001-eng.htm.

Statistics Canada. 2011. "Television Broadcasting" (November 1). http://www.statcan.gc.ca/daily-quotidien/111101/dq111101a-eng.htm.

Telefilm Canada. 2012. "Fostering Cultural Success. Accessed May 29, 2012. http://www.telefilm.ca/en/telefilm/corporate-publications/corporate-plan.

Vlessing, Etan. 2012a. "Few Netflix Canada Subscribers Cut the (Cable) Cord." *The Hollywood Reporter*, March 21.

———. 2012b. "Cord-cutting on the Rise." *Playback*, April 12.

2

Sound Recording and Radio: Intersections and Overlaps

Richard Sutherland

When Paul Audley published *Canada's Cultural Industries* in 1983, he remarked on the degree to which sound recording and radio broadcasting industries were "closely interrelated" (140). The basis of this interrelatedness was the arrangement whereby the sound recording industry supplied the bulk of radio broadcasters' programming, receiving in exchange the promotional benefits accruing from airplay. By the time Audley was writing, this symbiotic relationship had become thoroughly entrenched. Since the late 1950s, radio had been the primary vehicle for marketing sound recordings to the public. Disputes between radio and the sound recording industries over the relative value of programming and promotion were by no means uncommon, but the basic terms of the relationship were more or less unquestioned. This relationship was also the basis for, and ramified by, a number of government policies, such as Canadian content regulations for radio and the Sound Recording Development Program (SRDP). It was also a focal point for much of the copyright revision process up to the late 1990s.

Today, nearly thirty years after Audley's book, it is worth noting how much of the relationship remains intact. Amongst the proliferation of new media and the considerable reconfiguration of cultural industries, music still constitutes roughly three-quarters of Canadian radio programming (Copyright Board of Canada 2005, 21). Record companies continue to devote considerable resources to obtaining airplay on radio as a key strategy for reaching consumers, many of whom, according to broadcasters at least, still rely on radio for exposure to new music (CAB 2009, 2). Continued conflict between the two industries over the royalties

broadcasters pay to music industry copyright collectives, Canadian content levels, or Canadian content development funds is merely evidence
of the degree to which they remain concerned with one another. It is easy
to overlook, in discussions of newer commodity forms of music, such as
digital downloads or the renewed emphasis on blockbuster live shows,
that the music industry might be said to rely on radio more than ever. In
fact, as revenue from record sales has declined over the past decade, radio
is arguably the largest single source of music-derived revenue globally
(Resnikoff 2011). This is also the case in Canada, where radio's $1.6 bil-
lion in revenues is more than three times the estimate of those from the
sale of recordings.

Nonetheless, while there is much that continues to tie these industries
together, their development has been very different, particularly over the
past decade as each confronts the effects of digitization. One sign of change
is found in the way we refer to one of them. The two previous versions
of this book refer to the sound recording industry (Audley 1983; Dorland
1996). There were two reasons for this. First, it was often understood by
industry and by governments that sound recording was broadly emblematic of other aspects of the music industry, such as live performance and
music publishing. Second, this was the case because, in many respects,
sound recording actually was at the centre of the music industry—the
main source of revenue from music and the driver of revenues for other
music-related industries. However, in this edition, I choose to use the term
"music industry," largely because this has become the more accepted term.
For instance, the Canadian government's Sound Recording Development
Program was superseded in 2001 by the Canada Music Fund. Two of the
major industry groups in Canada have similarly altered their names. The
Canadian Independent Record Production Association (CIRPA) changed
its name to the Canadian Independent Music Association (CIMA) in
2010. More recently, the Canadian Recording Industry Association (CRIA)
became MusicCanada.[1] Although these may be largely cosmetic changes
in terms of the membership and mission of these two organizations, they
do reference the fact that the once dominant source of revenue (that is,
selling sound recordings to consumers) is in decline. Meanwhile, new
business models for marketing music continue to proliferate, but these are
in many instances unproven or unstable. On the other hand, commercial

---

1 Williamson and Cloonan (2007) have asserted that the use of the term "music industry" on the
part of the record label associations is also an attempt to co-opt other music-related industries,
such as live performance or music publishing, in a bid to become the primary spokespersons to
government, other industries, and the public.

radio, despite increased competition from some of the same technologies and business models, carries on more or less as it has since the 1920s: delivering audiences to advertisers. An examination of these industries in a Canadian context presents even more contrasts between them. Whereas a specifically Canadian music industry is becoming an increasingly difficult entity to identify or describe, Canadian radio broadcasting remains relatively distinct from its global counterpart, even as it operates more and more within larger Canadian media conglomerates. Therefore, I will consider these two industries in two different sections, providing an overview of their present circumstances and describing the paths that have brought them there. Government policy continues to play a prominent role in both industries, particularly in terms of the relations between the two, forming some of the most persistent bonds between them. As such, it is appropriate to deal with government policy for both industries in one section.

The Music Industry

As far as performers are concerned, Canadian musicians appear to be thriving. The 2011 Grammy Awards broadcast was in some respects a showcase for Canadian music. The iconic Canadian singer-songwriter Neil Young captured the first Grammy of his career with the Award for Best Rock Song. Newer artists such as Justin Bieber and Drake performed during the broadcast, although neither was able to parlay their several nominations into awards. It was Montreal-based Arcade Fire, however, that provided the evening's most dramatic development, unexpectedly winning the Grammy Award for Album of the Year for its 2010 release *The Suburbs* and giving an outstanding performance that closed the show. The global music industry's highest-profile event attested to the conspicuous success enjoyed by a number of Canadian musicians. Not only did their careers span nearly half a century, but also the presence of Canadian artists in the genres of teen pop and R&B would seem to indicate a greater stylistic diversity than this country has generally been noted for (Straw 1996, 114–115).

What all of this means for the Canadian music industry, however, is perhaps less clear. None of the above acts is signed to a Canadian label (or even to the Canadian subsidiary of a multinational). Each of the artists found international success simultaneously with Canadian recognition. Notwithstanding the intensity of some Canadians' identification with these artists, the development of their careers has very little to do with a Canadian-based industry. This should not surprise us. We are used to Canadian television and film stars whose careers play out almost entirely in the larger markets and global entertainment hubs of the United States.

Neil Young himself is evidence that the disconnection between the two is nothing new for Canadian music. The point is that it is not necessary to have a Canadian music industry in order to have Canadian artists. It would be equally true to say that Canadian artists are not a necessary requirement for a music industry in Canada. It remains the case that the music industry in Canada is engaged primarily in selling internationally successful artists to Canadians. According to Statistics Canada (2009b, 5), in 2007 recordings by Canadian artists accounted for just below 26 per cent of revenues for recording companies operating in Canada, and this is a relatively high proportion, historically (Wikström, 2009, 72). There is no doubt that Canada is, and has been for a long time, integrated into the global music industry.

An Industry in Crisis?

From either a Canadian or a global perspective, the music industry appears to be in a period of considerable turmoil—by many accounts, this is a result of the diffusion of digital technologies over the past two decades. Industry groups such as the International Federation of Phonographic Industries (IFPI) point to the decline in sales of recordings and the concomitant decline in revenues for the industry over the past decade, blaming the widespread exchange of music files over the Internet (IFPI 2011). IFPI speaks of a "crisis" as the industry sheds revenue, investment, and jobs (3) and, clearly, sales of recordings have declined precipitously over the past decade. In Canada, for instance, in 2010 a total of 31.4 million albums were sold (FYI Editor 2011), down from 47.7 million in 2006 (Nielsen Music 2008)—a drop of 34 per cent in five years, coming on top of several years of decline. Given the fact that album prices have declined over the same period, the effect on revenues can be easily understood. However, despite these numbers, some scholars have argued that the picture is not as bleak as it appears. For instance, Winseck (2011) has offered some grounds for believing that music-derived revenues may actually have increased over this period. He shows that while revenues from sound recordings have certainly declined, revenues for music publishing have remained relatively stable, and those derived from live performance or from the Internet and mobile phone market have actually increased substantially.

There is also dispute over precisely what has caused recording sales to decline. The most frequently cited cause is consumers accessing unauthorized digital copies of music over the Internet, primarily via peer-to-peer networks. While it would be surprising if the easy availability of free music had no impact on music sales, it is far from conclusive that it has wrought as much damage as the music industry claims. It may also be that the decline in recording sales is linked, at least in part, to increases in sales of other entertainment products, such as video and computer games or the variety of electronic devices, such as iPods or cell phones (and the fees for using them), and even concert admissions, all of which compete with sound recordings for consumers' dollars (Sutherland and Straw 2007, 144–145). The actual effect that downloading has on music sales is difficult to assess, with different studies yielding different results. A 2007 study for Industry Canada found, contrary to much of the existing research, that file sharing actually tended to increase music purchases (Anderson and Frenz). In 2011, the Canadian Intellectual Property Council, an industry interest group, estimated that seven million Canadians were downloading unauthorized music files from the Internet, costing the recording industry $1.26 billion annually (26). Regardless of these ambiguities, the music industry has fought a number of high-profile legal battles over the last decade against such services, including Napster, LimeWire, and The Pirate Bay. As well, the industry has launched lawsuits against individuals accused of uploading or downloading these files with varying degrees of success. In Canada, these lawsuits foundered on technicalities (see Longford 2007) with the result that Canada's laws in this area were, until very recently, ambiguous.

The music industry in Canada, as well as globally, is highly concentrated. In 2011, 80 per cent of the Canadian market was in the hands of four multinational record labels. With the recent sale of EMI's record label to Universal Music Group, the market is now even more concentrated. The recording and music publishing sectors are also relatively separate from other entertainment industries in terms of ownership. A decade ago, the five major labels were nearly all divisions of global media conglomerates. Today, this is the case only for Sony Music. For the most part, this separation holds true for the several hundred independent labels operating in Canada, most of which are extremely small enterprises. Even the larger independent labels, such as Arts & Crafts Records or Last Gang Records, are stand-alone companies. The few exceptions to this isolation are found in Quebec, where some of the larger independents, particularly Distribution Select (part of media giant Quebecor), have ties to other sectors of the entertainment and media industry.

Declining Physical Sales

For much of the twentieth century the Canadian sound recording industry was a manufacturing industry, concerned with pressing and distributing copies of recordings under license from foreign-owned companies. Companies such as Compo (the vestiges of sound recording pioneer Emil Berliner's empire) and Quality took advantage of the high tariffs on manufactured goods to provide a reliable if small income to licensors such as Warner Brothers Records or Decca. This changed as the major labels opened their own offices in Canada. This development coincided, roughly, with the introduction of Canadian content, but had less to do with the possibility of airplay for Canadian acts than with the realization that setting up a wholly owned branch office subsidiary could substantially increase revenues from the Canadian market (Cornyn, 189). This was part of a wider strategy on the part of the major labels to take control of distribution in North America as a whole, but it had substantial benefits for the entire Canadian music industry. These networks offered Canadian-owned labels an effective means of marketing their own recordings across Canada through distribution deals with the major labels.

From the 1970s to 2000, the major labels remained in firm control of the Canadian market both through their domination of sales and their function as distributors for most significant independent labels. Even when the tariff on imports of sound recordings from the United States was removed in 1997, the major labels maintained their Canadian distribution operations. It was only between 2003 and 2007 that they outsourced the physical distribution functions that had been the basis of their control, faced with the cost of maintaining several large networks in competition with one another, especially in the face of a decline in sales of physical copies of albums. Physical distribution for all of the industry in Canada is now managed by the Markham, Ontario-based disc manufacturer Cinram (Leblanc 2004, 2007). This development is indicative of a shift in business strategy from manufacturing to marketing proficiency and copyright management. As table 2.1 shows, the major labels still operate in Canada and still account for roughly 80 per cent of recordings sold here, but they preside over a smaller market and their physical footprint is now much smaller (Nielsen 2011).

Declining sales of physical recordings have had the most visible impact on Canadian music retailers. To say the least, the past decade has not been kind to them. In 2001, Canada's best known music retail chain, Sam the Record Man, went into receivership. It was followed in 2007 by Music World and in 2008 by A&B Sound. With the demise of these stores,

Table 2.1 Market Share of Major Labels in Canada in 2011 (2010)

Label	% Share Current	% Share Catalogue
UMG	40.2 (39.12)	35.98 (35.81)
Sony	17.86 (19.37)	19.97 (21.98)
Warner	11.33 (13.69)	16.24 (15.38)
EMI	10.1 (8.5)	10.7 (11.7)

Source: Nielsen Company (FYI 2011).

Canada lost its last domestically owned music outlets. Nor does the future look especially bright for the sole remaining national chain, HMV. In June 2011, HMV's British parent, working under a heavy debt load, sold its Canadian operations to Hilco, a UK-based restructuring specialist, for just $3.2 million ("HMV Canada Sold for $3.2 Million" 2011). At the point of sale, the new owners intended to close some of the chain's 121 stores and to increase its online presence. HMV's retreat from the market leaves music retail in the hands of big-box department and electronics stores, alongside a dwindling number of independent retailers.

Unlike retailers, record labels have derived some benefit from a legitimate online market for recordings; however, dominance has shifted to other businesses, notably Apple, whose iTunes store is now the largest music retailer in the world. Music may be a relatively minor part of Apple's business, but its dominance of the market in legal downloads (in the region of 70 per cent) gives it considerable power over the industry and helps to market the electronic devices that are its main source of revenue. Other companies that are a part of this new online distribution infrastructure include Internet service providers and search engines such as Google. All could be said to derive at least part of their revenues from consumers accessing music, but little or none of this income finds its way back to the businesses that previously made up (and still, at least notionally, define) the music industry, including record labels and music publishers, nor, indeed to artists and songwriters.

The Developing Digital Market

Digital music sales in Canada are growing rapidly, although this has been more than offset by the declines in physical sales. Digital copies now constitute 19 per cent of total albums sold (FYI Editor 2011), in a market that is still defined by single-song downloads. The development of this online market has been marked by the creation of dozens of services all aimed at finding ways to profitably sell music to consumers within the law, encompassing a wide array of business models and with varying degrees of success. During much of the past decade, the legitimate online market for music appeared to be slow to develop in Canada, iTunes arrived in Canada three years after its launch in the United States, and many other services seemed to be slow in making their way to Canada. The Canadian music industry placed the blame on our lack of adequate copyright laws (Henderson 2006). Others, however, have suggested that it is the music industry itself that has been the major obstacle. iTunes was delayed at least in part by protracted negotiations with Canadian rights holders (Dixon and Blackwell 2003). Pandora publicly blamed the record labels for demanding a disproportionate amount of the revenue and opted not to enter the Canadian market at all, while other services have pointed to the relatively small size of the market as a deterrent (Lambert 2010). In fact, although Canada's digital market was somewhat slow to get started, a vertiginous growth rate since 2007 has brought the market to levels that are not very far off those of the global market, this in advance of copyright revision. While many of the most prominent of these services are not available in Canada, we can find examples of most of the kinds of business models active in the Canadian market. Following Patrik Wikström (2009), these can be grouped into several different varieties (see table 2.2).

In addition to this list there are countless other sources of free and legal music on the Web posted by artists or labels themselves, either through their own websites or via social media sites, such as Facebook or Myspace.

Licensing and Music Publishing

Faced with a decline overall in sales of recordings, the industry is becoming more reliant on other sources of revenue from licensing and royalties generated by industrial users of recorded music. These include film and broadcast media, the electronic gaming industry, and advertisers and a host of other businesses that use music as part of their operations. Statistics Canada put the estimate for the recording industry's revenues from licensing and royalties at about $40.2 million in 2007 (Statistics Canada 2009b, 8). In the same year, the Canadian music publishing

Table 2.2 Digital Music Services Available in Canada

Single Song Download	Subscription – Limited Quota	Membership – Unlimited (streaming)	Ad-based
iTunes Canada (Apple)	Emusic	Rhapsody	Slacker Radio
Zik (Quebecor)	Motime	Rdio	YouTube (Google)
PureTracks		Slacker Radio	Vevo Canada (CTV, major labels), video only
Ur Music (Rogers)			
Bell Mobility			
TELUS Mobility			
7 Digital			

industry, which derives most of its income from various forms of licensing, saw revenues of $122.9 million, with this number rising to $141.7 million in 2008 (2). Perhaps the largest single source of such revenue is through performing rights, which are levied on all forms of public performance of music in Canada by SOCAN (the Society of Composers, Authors and Music Publishers of Canada), on behalf of songwriters and music publishers. In 2010, SOCAN collected $229.5 million in Canada (SOCAN 2011a). This accounts for much of the revenue for music publishers. There are forty-five different tariffs collected by SOCAN for uses that range from cable television to background music at skating rinks (SOCAN 2011b). A smaller but growing source of licensing revenue comes from neighbouring rights, collected by Re:Sound on behalf of record companies and recording artists for the public use of recorded music.

Live Performance

The live music sector is another important component of the music industry, with revenues in 2008 that were at about two-thirds of what was derived from the sale of recordings. Embracing a variety of forms from nightclub performances to concerts and festivals, it is a difficult field to document. An exploratory study conducted for Canadian Heritage in 2004 used SOCAN's tariffs on

Live Music and Popular Music Concerts as a means of charting growth, while noting that these were not always distinguished between higher fees for artists or increased attendance (Department of Canadian Heritage 2004, 5). Using SOCAN's methodology, we can see that after a substantial increase of 6 per cent to just over $17 million in 2007 (SOCAN 2008), concert revenue did not grow by very much and even declined to about $16 million in 2010, which the society attributed to fewer big-name tours over that year (SOCAN 2011a, 3). Statistics Canada figures on performing arts revenue for 2008 and 2009 show that the total revenue for the sector (including both not-for-profit and for-profit performers) fell from $416.9 million to $385.1 million (2009a, 2). This suggests that live performance, while significant as a source of revenue, is by no means experiencing the enormous growth some have suggested. In fact, it appears the North American live music market peaked in 2008 and has been in decline since then, perhaps a victim of the economic downturn ("2010 By the Numbers (So Far)" 2010; Smith 2009).

Policy
Copyright

As the largest industrial user of music, radio generates considerable royalties for the music industry. These are set by the Copyright Board of Canada. The largest set of royalties is in the form of the performing rights paid to SOCAN. In 2010, SOCAN collected roughly $56 million from Canadian radio stations (SOCAN 2011a, 3) based on a tariff ranging from 1.5 to 4.4 per cent of advertising revenues (Copyright Board of Canada 2008, 5). In the last major round of copyright revision in Canada in 1997, these rights were supplemented by two new rights: a home taping right in the form of a tariff on blank recording media (collected by the Private Copying Collective of Canada on behalf of rights holders) and neighbouring rights (the rights payable by broadcasters and other users of recorded music to record companies and performers). The introduction of neighbouring rights resulted in further payments by radio stations for the music they program. This right is collected and administered by the copyright Re:Sound (formerly the Neighbouring Rights Collective of Canada) on a tariff that ranges from 0.75 to 2.1 per cent of advertising revenue (Copyright Board of Canada 2008, 5).

Over the past decade copyright has emerged as perhaps the music industry's number one concern as it begins to address the growth in unauthorized downloads among consumers. Governments have responded with four attempts since 2004 at passing copyright revisions that would address the technological developments of the past two decades. This became a particular priority for Stephen Harper's Conservative government because the lack of an up-to-date copyright regime of the past many years has been a major irritant in trade relations with the United States. Finally, Bill C-11, An Act to Amend the Copyright Act, was passed in June 2012. This bill establishes the right of record labels to prevent both uploading and downloading of unauthorized music files. It also sets a maximum fine of $5,000 as the penalty to consumers who engage in these activities (Lithwick 2010) and specifies what rights consumers do enjoy regarding copying music they have acquired legitimately.

Subsidy and Contribution Programs

Radio broadcasters have also played a critical role in the establishment of subsidy programs for Canada's music industry, participating in the creation of FACTOR (the Foundation to Assist Canadian Talent on Record) in 1982. FACTOR eventually became the vehicle for the government programs created under the Sound Recording Development Program (SRDP) in 1986. The SRDP was subsequently incorporated into the expanded program of the Canada Music Fund in 2001, whose budget reached $9.85 million in 2009 (Department of Canadian Heritage 2009). These programs were aimed exclusively at Canadian-owned companies and within that sector brought in a certain level of stability to some companies; as well, they allowed emerging artists to record and market their music. The radio industry was involved in the creation of another funding body for Canadian music, the Radio StarMaker Fund, created in 2001. Three per cent of the value of all ownership transactions in the commercial radio sector was to go to this fund, with another 2 per cent going to FACTOR and a further 1 per cent to these or other initiatives at the licensee's discretion. Radio stations have made contributions under Canadian Content Development commitments both to FACTOR and to a host of smaller local or station-run initiatives. The total value of these contributions to all programs and initiatives was $46.1 million in 2010 (CRTC 2011, 54).

Content Regulations

Canadian content regulations for radio were for a long time the flagship cultural policy for the Canadian music industry, since their introduction by the CRTC in 1971. Rightly or wrongly they have often been accorded a fundamental role in establishing a Canadian music industry. Their durability, at least in broadcast radio, is itself notable and perhaps a testament to this legacy. Although Canadian content was once accompanied by an elaborate set of programming guidelines for FM radio, there are now few remaining content regulations for Canadian radio stations. The last vestige of the CRTC's FM radio policy, the hit/non-hit ratio that limited the number of "hit" songs FM radio could play, was abandoned in 2009. Stations are free to define and switch formats without permission, although such decisions can be reviewed at the time of licensing renewal.

The limitations of Canadian content began to emerge even in the early 1980s as the policy was clearly insufficient to guarantee a reasonable market share for Canadian artists. In the late 1990s the concern over the lack of access to the airwaves of emerging and local artists began to cause some to look at restructuring Canadian content regulations as a way to address the problem. The 1998 regulations, however, did nothing to resolve this, although they did raise the level of Canadian content to 35 per cent. The call for a revamped quota system was repeated more strongly at the CRTC's review of radio policy in 2006 but did not result in any substantive changes to the regime, although stations were encouraged to make voluntary commitments to set aside programming for emerging Canadian artists. Another limitation on Canadian content has been its limited application to other audio media. The CRTC chose not to impose any content regulations on Internet-based services, and they have only been applied to satellite radio in a very limited and partial manner. There is a similar case with French-language vocal regulations. While these remain in place for francophone stations, at the level of 65 per cent of vocal selections, they have not applied to other audio media.

Ownership Regulations
In 1998, as part of its new regulations for radio, the CRTC introduced a key change in the rules limiting the number of stations within a single market for a single owner (CRTC 1998, 38). Previously, radio stations had been limited to one AM and one FM station in any given market, but the regulator expanded this to two AM and two FM stations. This measure was taken in response to the radio industry's concerns about its stagnant revenues in the early 1990s and the prospect of increased competition from new media in the coming years. In exchange for easing ownership restrictions, the Commission required that 6 per cent of the total cost of the transaction was to be assigned to Canadian Talent Development (as it was then called), as referred to above. The 2006 hearings on radio policy did not specify any changes in the ownership limits in a market but suggested instead that all applications for ownership transfer or new licenses would be assessed on a case-by-case basis (CRTC 2006, 158–206). Furthermore, the Commission reaffirmed the principle that any award of a new license or transfer in ownership would now incur a fee to be assigned to Canadian talent development.

Radio

As radio broadcasting nears the end of its first century in Canada, it remains a relatively healthy industry. Having survived the arrival of television some sixty-odd years ago, radio faces new competition from a plethora of new technologies that offer listeners unprecedented choice for audio entertainment and information. Yet revenues are fairly stable for commercial radio. In 2010, they rose to $1.55 billion, up 2.9 per cent, having resumed a pattern of steady growth after a year in which they had declined by just over 5 per cent (CRTC 2011, 18). As of 2011, there are, in total, 1,208 audio services operating in Canada, thirteen fewer than in 2010 (35). Of this total, 733 are commercial stations—543 FM stations and 143 AM stations (36). (The remainder are networks or digital licenses.) The number of AM stations has been on a steady decline for some years, but much of this has occurred as broadcasters convert their licenses to the superior FM signal. Five hundred and fifty-eight of these commercial stations broadcast in English, 104 in French, and a further 24 in other languages (36).

The Canadian radio industry is also relatively concentrated, with the top five broadcast chains accounting for 67 per cent of the revenue for the sector. Collectively, they account for 289 stations across Canada. In addition, these chains are themselves integrated into larger media conglomerates, with considerable holdings in other media, including television, cable, publishing, and telecommunications. Both the number of stations and their ownership by large media groups is part of a relatively consistent trend toward integration in the Canadian media landscape. For radio, this trend accelerated in the past two decades as a result of the changes in ownership regulations discussed in the policy section above.

Table 2.3 The Top Five Radio Broadcast Chains

Company	Share %
Astral	21%
Corus	16%
Rogers	13%
BCE (CTV-Globemedia)	10%
Newcap	7%

Source: CRTC (2011).

While financially, Canadian radio appears to be relatively stable, it does nevertheless face some key challenges going forward. Looking at listening statistics, one can see a decline in the number of hours Canadians tune in to radio. In 2010, Canadians listened to radio, on average, 17.6 hours per week (CRTC 2011). This was down from 19.1 hours in 2005 (Statistics Canada 2007a). But the decline has been particularly pronounced in younger demographics. Already by 2007 teens listened to radio only 7.2 hours per week (Statistics Canada 2007b), and that number itself was a sharp drop from 2001 figures (Statistics Canada 2007a). The reasons for this decline may vary, but it is partially due to the numerous alternatives for music listening available through the Internet or on portable devices such as iPods. Radio's long reign as the primary source of free music is challenged by Internet radio, streaming, and downloading. Satellite radio is also to some extent in competition with conventional broadcast radio, although its cost limits this to some extent. However, whatever the cause, if these trends continue as they are, radio might well find its audience, and hence its revenues, increasingly limited in the coming years.

Internet Broadcasting and Podcasting

Internet broadcasting presents radio stations with opportunities as well as threats. In terms of the former, it provides them with another relatively low-cost means of reaching audiences outside of their market areas, although this has little impact on their advertising revenues, which still depend on conventional ratings. A large number of commercial radio stations engage in some form of Internet broadcasting, using it not only to rebroadcast but also to supplement their offerings to audiences. However, the extent to which the Internet presents such opportunities is more than balanced by competition for listeners outside the market. This includes some of the services, such as Rdio, mentioned in table 2.2. But it also includes stations from other markets, not only in Canada but around the world. In the mobile market, apps such as TuneInRadio offer listeners easy access to streaming audio from hundreds of broadcasters worldwide. A 2009 Statistics Canada survey of Internet use by Canadians indicated that 32 per cent of Internet users listened to radio over the Internet and 47 per cent used the Internet to obtain music, either free or paid (Statistics Canada 2010).

Satellite Radio

Satellite radio has enjoyed a relative measure of success in Canada since its introduction in 2005, but this has not been without its difficulties. Satellite radio allows subscribers access to 150 channels that play a wide array of tightly formatted programming. In the US two services emerged in the late 1990s: Sirius and XM Satellite Radio. The American companies needed Canadian partners to offer their services here and some interesting ownership configurations resulted. Canadian Satellite Radio (CSR) was a partnership between XM Radio and Toronto businessman John Bitove. Sirius Canada involved private radio chain Standard Broadcasting (now owned by Astral Media), as well as a stake from the CBC. The CRTC approved both Sirius Canada and Canadian Satellite Radio in early 2005. This decision was appealed and sent back to Cabinet, which approved the original decision in September 2005, after some concessions in the area of French and Canadian content were made by both services ("Satellite Radio Given the Okay" 2005). Even so, the content requirements were far below that of conventional radio, averaging out at less than 10 per cent across the system.

Satellite radio was somewhat slow to take off in Canada, with Sirius Canada and CSR incurring heavy losses, as did their parent companies in the United States. In fact, so deep were the losses of the American companies that they were both in danger of going bankrupt, leading them to merge in 2008. In June 2011, Sirius Canada and XM Satellite Radio Canada followed suit. If the Canadian merger seemed more or less inevitable, it was delayed because the more complicated ownership structure of the Canadian operations caused some dispute over the relative value of each company in the newly merged entity (Krashinsky 2010). The newly merged entity has about 1.8 million subscribers and revenues of $200 million annually (XM Satellite Radio Canada 2011).

Digital Broadcasting

Radio has thus far been less touched by digital broadcasting than has television. In 1995 the CRTC established an elaborate plan for the transition to digital radio (CRTC 1995), but there was very little uptake on the part of stations. As of 2011, there are only forty-one digital radio licenses in Canada, most of which rebroadcast signals from conventional radio (CRTC 2011, 36). In 2006 the CRTC itself noted that the development of digital broadcast radio had stalled; not much has changed over the last five years. One significant factor, perhaps, is that the automobile industry has put its energy into satellite radio rather than Digital Broadcast Radio (CRTC 2006, 6). Currently, there is no plan to eliminate conventional radio broadcast as happened with television in 2011.

Public Broadcasting: CBC/SRC

The Canadian Broadcasting Corporation celebrated its seventy-fifth anniversary in 2011, which marked the date of its conversion to a Crown corporation (it had earlier been the Canadian Radio Broadcasting Commission, created in 1932). At the time, the CBC played an absolutely central role in Canadian radio as the sole network, as well as the regulator for the industry. This latter role changed in 1958, with the creation of the Board of Broadcast Governors (BBG), and today the CBC must account to the BBG's successor, the Canadian Radio-television and Telecommunications Commission (CRTC), as any other broadcaster does. In terms of its popularity, the CBC is by no means the dominant force it once was, yet it remains an important element in Canadian radio with its two English-language broadcast networks, Radio 1 and Radio 2, maintaining a 14.7 per cent share of the English radio audience in Canada (CBC 2011, 31), while its equivalent French services account for 19.5 per cent of the francophone audience (33). Like its commercial counterparts, CBC Radio has also embraced new technologies, and its programming is available not only over the air but via Internet radio, podcasting, and even on satellite radio.

Programming over CBC Radio 1 remains a blend of national programming, alongside local and regional offerings, combining news and entertainment. Radio 2 is almost entirely network programming, primarily music, encompassing a variety of genres. The service was revamped in 2008, reducing its classical music programming in favour of more popular genres, a move which also generated some controversy at the time ("Cross Canada Protests..." 2008). The English-language service also runs Radio 3, an Internet broadcaster specializing in popular music by emerging Canadian artists. The French-language side also runs two broadcast networks, Première Chaîne (largely information-based programming) and Espace Musique, and its Internet station Bande à Part is focused on new and emerging francophone artists.

The CBC does not report precisely how much of the $1.1 billion the corporation receives in public funding is spent on radio. Nonetheless, it would be fair to say that the radio side of CBC relies more or less entirely on government support, as the service has not run advertising since 1974. To some extent this has insulated CBC Radio from some of the political pressure felt by the television side of the corporation, which competes with private broadcasters for advertising dollars as well as audiences. Nevertheless, the decline in television advertising revenues and pressure on government budgets led to speculation in 2009 that the radio service would start running paid advertising (Ha and Bradshaw 2009). At the time that notion was very quickly quashed, but the idea was revived again in 2012 in the wake of further budget cuts. Yet the radio service does have some more commercial dealings through its involvement with satellite radio. CBC is one of the partners in Sirius Canada, which carries its radio services across North America, and it retains a 15 per cent stake in the newly merged Sirius-XM Canada (CBC 2011, 103).

Canadian radio broadcasting is less integrated into a global industry than is sound recording or other music industries. Canadian radio stations can and do follow trends in programming and marketing from abroad, particularly the United States, but beyond this the Canadian industry's interaction with its global counterpart is fairly limited. Ownership regulations may have been relaxed over the past couple of decades, but it remains the case that majority or controlling interest in broadcast properties must be Canadian. Regulation of the sector is an exercise in national sovereignty over the airwaves. In this respect radio seems somewhat exceptional, facing, as it does, Canadian content regulations that for other sound media are either much lower or even non-existent.

Conclusion: Technology, Convergence, and Globalization

Both the Canadian radio and music industries have faced challenges from new media technologies over the past decade. On the face of it, radio appears to be less affected by these changes in terms of its overall financial health and, perhaps more importantly, in terms of its basic business model. Annual revenues of over $1.5 billion likely far outstrip the collective performance of the various online music services that have entered the Canadian market over the past decade or so, although the decline in listeners make radio's long-term future somewhat murkier. The music industry, on the other hand, has seen a much more difficult transition thus far. Winseck's (2011) claims that music-derived revenues have increased depends on a fundamental reassessment of the kinds of enterprises that make up the music industry, as telecommunications and technology companies play a much more central role. Music continues to be an attractive commodity, but matters of who gets paid for it, and how, are far from settled.

Both sound industries operate not only in relation to one another but increasingly in concert with other media industries as well. For radio this has meant its integration into larger Canadian media conglomerates. Broadcast radio provides just one component in a coordinated strategy for content and advertising across various platforms. For the music industry, convergence has meant ceding control over distribution to other companies and, with this control, the future direction of the industry in subordination to the strategies of technology companies that will make the key decisions over the way music is marketed to consumers. The gradual shift to licensing and royalty revenues suggests a future in which music is a value-added service, rather than a consumer good in its own right.

Where we see the strongest divergence between the two industries is in terms of their relations to global media markets. The continued difficulties facing emerging and local artists are not easily positioned as issues of nationality or domination by foreign interests. Instead, they look a lot like the problems facing emerging and local artists everywhere—the difficulty lies in gaining the attention of consumers in a crowded field of competitors. For the Canadian industry, concerns over the access of Canadian musicians and producers to their own market seem less compelling in the face of the conspicuous global success of many Canadians. Canadian-based music business such as Nettwerk and Arts & Crafts are engaged with Canadian and foreign artists and pursue interests outside of the Canadian market. With this in mind, defining a Canadian music industry as a discrete entity is increasingly problematic. The degree to which this is still possible is largely the result of a global copyright regime still reliant on national sovereignty to define territories. Radio broadcasting is still far more defined by its nationality. Again, this is a function of sovereignty. The regime that asserts national control over the broadcast spectrum and insists on Canadian ownership and control of media properties has meant that Canadian radio is largely insulated from many of the consequences of globalization. The degree to which the two industries continue to intersect in Canada is based largely on this reality.

References

"2010 By the Numbers (So Far)." 2010. *Pollstar*. The Concert Hotwire (July 9). Accessed September 9, 2011. http://www.pollstar.com/blogs/news/archive/2010/07/09/731238.aspx.

Andersen, B., and Frenz, M. 2007. *The Impact of Music Downloads and P2P File-Sharing on the Purchase of Music: A Study for Industry Canada*. Birkbeck:, University of London, Faculty of Management.

Audley, P. 1983. *Canada's Cultural Industries: Broadcasting, Publishing, Records and Film*. Toronto: James Lorimer & Company, Publishers.

CAB (Canadian Association of Broadcasters). 2009. *Submission of the Canadian Association of Broadcasters: Copyright Consultations 2009*. Ottawa: CAB.

CBC (Canadian Broadcasting Corporation). 2011. *Annual Report*. Ottawa: CBC.

CIPC (Canadian Intellectual Property Council). 2011. *The True Price of Peer to Peer File Sharing*. Ottawa: CIPC. Accessed October 28, 2011. www.musiccanada.com/Assets/News/The%20True%20Price%20of%20Peer%20to%20Peer%20File%20Sharing.pdf.

CRTC (Canadian Radio-television and Telecommunications Commission). 1995. *Public Notice 1995-60: A Review of Certain Matters Concerning Radio*. Ottawa: CRTC.

———. 1998. *Public Notice 1998–41: Commercial Radio Policy*. Ottawa: CRTC.

———. 2006. *Public Notice PN 2006-160: Digital Radio Policy*. Ottawa: CRTC.

———. 2011. *Communications Monitoring Report, 2010*. Ottawa: CRTC.

Copyright Board of Canada. 2005. *Statement of Royalties to Be Collected by SOCAN and NRCC in Respect of Commercial Radio for the Years 2003 to 2007*. Ottawa: CBC.

———. 2008. *Statement of Royalties To Be Collected by SOCAN and NRCC in Respect of Commercial Radio*. Ottawa: CBC.

Cornyn, S., with P. Scanlon. 2002. *Exploding: The Highs, Hits, Hype, Heroes and Hustlers of the Warner Music Group*. New York: Harper Entertainment.

"Cross Canada Protests Decry CBC Radio Changes, Orchestra's End." 2008. CBC News: Arts & Entertainment. Accessed September 30, 2011. http://www.cbc.ca/news/arts/music/story/2008/04/11/radio2-protests.html.

Department of Canadian Heritage. 2004. "Overview of the Live Music, Festival and Concert Industry in Canada." Accessed July 15, 2007. http://www.pch.gc.ca/pc-ch/pubs/concerts/cont_e.cfm.

———. 2009. "Government of Canada Renews Canada Music Fund and Increases Investment in Digital and International Market Development" (July 31). Accessed September 2, 2009. http://www.pch.gc.ca/eng/1294862453819/1294862453821.

Dixon, G., and R. Blackwell. 2003. "Puretracks Launches Legal Way to Download." *Globe and Mail*, October 15, R3.

Dorland, M. 1996. *The Cultural Industries in Canada: Problems, Policies and Prospects*. Toronto: James Lorimer & Company, Publishers.

FYI Editor. 2011. "Canadian Music Market Sales Report, 2010." FYI Music News, April 5. Accessed September 21, 2011. http://fyimusic.ca/industry-news/canadian-music-market-sales-report-2010.

Ha, T.T., and J. Bradshaw. 2009. "CBC Won't Air Ads on Radio Programming." *Globe and Mail*, March 18.

Henderson, Graham. 2006. "Beyond Deadwood: Canadian Law and the Development of a Digital Marketplace." Speech to Canadian Club (electronic version), Toronto, ON, May 1.

"HMV Canada Sold for $3.2 Million." 2011. *CBC News*, June 27. Accessed July 12, 2011. http://www.cbc.ca/news/arts/story/2011/06/27/hmv-canada-sale-hilco.html.

IFPI (International Federation of Phonographic Industries). 2011. *IFPI Digital Music Report 2011: Music at the Touch of a Button*. London: IFPI.

Krashinsky, S. 2010. "Sirius, XM Canada Reach Merger Deal." *Globe and Mail*, November 25.

Lambert, Steve. 2010. "Music Streaming Services Reject Canada." *Globe and Mail*, September 23.

Leblanc, L. 2004. "Cinram Adds UNI." *Billboard*, February 14. Accessed July 9, 2007. http://www.billboard.biz/bbbiz/search/article_display.jsp?vnu_content_id=2085955.

———. 2007. "Canada's Big Chill." *Billboard*, March 10. Accessed July 9, 2007. http://www.billboard.biz/bbbiz/search/article_display.jsp?vnu_content_id=1003552654.

Lithwick, Dara. 2010. *Legislative Summary of Bill C-32: An Act to Amend the Copyright Act* (July 20). Accessed September 2, 2010. http://www.parl.gc.ca/About/Parliament/LegislativeSummaries/bills_ls.asp?Language=E&ls=C32&Mode=1&Parl=40&Ses=3&source=library_prb.

Longford, Graham. 2007. "Download This!: Contesting Digital rights in a Global Era: The Case of Music Downloading in Canada." In *How Canadians Communicate* Vol. 2, edited by David Taras, Frits Pennekoek, and Maria Bakardjieva, 195–216. Calgary: University of Calgary Press.

Nielsen Music. 2008. "Year End Music Industry Report for Canada." *Reuters – Edition US*, January 4. Accessed September 22, 2011. http://www.reuters.com/article/2008/01/04/idUS214743+04-Jan-2008+BW20080104.

Resnikoff, P. 2011. "How Much Is the Music Industry Really Worth? Try $168 billion…" *Digital Music News*, September 7. Accessed September 7, 2011. http://www.digitalmusicnews.com/stories/090611industry.

"Satellite Radio Given the Okay." 2005. *Globe and Mail*, September 9.

Smith, Ethan. 2009. "Concert Industry Bucks Recession." *The Wall Street Journal*, January 1.

SOCAN. 2008. *SOCAN 2007 Finanical Report*. Toronto.

SOCAN. 2011a. *SOCAN 2010 Financial Report*. Toronto.

SOCAN. 2011b. SOCAN Tariffs. Accessed September 30, 2011. http://www.socan.ca/jsp/en/resources/tariffs.jsp.

Statistics Canada. 2007a. "Commercial Radio Listening Stabilizes." Accessed August 27, 2009. http://www41.statcan.ca/2007/3955/ceb3955_003-eng.htm.

Statistics Canada. 2007b. *Service Bulletin: Radio Listening: Data Tables*. Ottawa: Statistics Canada.

Statistics Canada. 2009a. *Service Bulletin: Performing Arts*. Ottawa: Statistics Canada.

Statistics Canada. 2009b. *Service Bulletin: Sound Recording and Music Publishing*. Ottawa: Statistics Canada.

Statistics Canada. 2010. "Canadian Internet Use Survey." *The Daily*, May 10. [Electronic Edition]. Ottawa: Statistics Canada.

Straw, W. 1996. "Sound Recording." In *The Cultural Industries in Canada: Problems, Policies and Prospects*, edited by M. Dorland, 95–117. Toronto: James Lorimer & Company, Publishers.

Sutherland, R., and W. Straw. 2007. "Canadian Music Industry at a Crossroads." In *How Canadians Communicate*, vol. 2, edited by David Taras, Frits Pannekoek, and Maria Bakardjieva, 141–165. Calgary: University of Calgary Press.

Wikström, P. 2009. *The Music Industry*. Cambridge: Polity.

Williamson, J., and M. Cloonan. 2007. "Rethinking the Music Industry" *Popular Music* 26 (2): 305–322.

Winseck, D. 2011. "Introductory Essay: The Political Economies of Media and the Transformation of the Global Media Industry." In *The Political Economies of Media and the Transformation of the Global Media Industry*, edited by D. Winseck and D.Y. Jin. London: Bloomsbury Academic. Accessed September 19, 2011. http://www.bloomsburyacademic.com/view/PoliticalEconomiesMedia_9781849664264/chapter-ba-9781849664264-chapter-001.xml.

XM Satellite Radio Canada. 2011. "Merger News: Sirius and XM Canada Complete Merger" (June 21). Accessed September 20, 2011. http://www.xmradio.ca/about/merger.cfm.

3

Newspapers and Magazines: Of Crows and Finches

Christopher Dornan

Janine Melnitz:	*You're very handy, I can tell. I bet you like to read a lot, too.*
Dr. Egon Spengler:	*Print is dead.*
	– *Ghostbusters* (1984)

Newspapers

Let us begin with a question to which we know the answer: In the second decade of the twenty-first century, is there a bright twenty-one-year-old anywhere in Canada, casting about for a career in the creative arts or media industries, who thinks, "Newspapers—that's where the future lies!"

When one has lost the interest of young people, one is in the mortuary business. For all that they trade in narrative and the ongoing documentation of contemporary affairs, the narrative *about* Canadian newspapers, the documentation of *their* performance and prospects, has been a chronicle of decline for forty years: loss of market, loss of advertising, loss of relevance, loss of confidence.

This is true of the newspaper industry throughout the Western democracies. But it is all the more significant considering the cultural place newspapers once enjoyed in Canada. Other Canadian media industries have been buffeted by the same turbulence—new sources of content and entertainment, the fragmentation of traditional markets, the migration of advertising, new technologies of content creation and distribution, new models of compensation—but these industries have prospered over the past decades, certainly relative to what existed in the 1970s. From the late twentieth century and into the early twenty-first, Canadian film and television production, Canadian popular music, and Canadian fiction have become

Table 3.1 Average Published Per Day Circulation of Top 20 Canadian Newspapers

Newspaper	1995	2010	% change
Globe and Mail	314,972	317,781	+0.9
Toronto Star	519,070	292,003	-43.7
Journal de Montréal	287,986	209,986	-27.1
La Presse	212,527	182,375	-14.2
Vancouver Sun	210,964	178,672	-15.3
Vancouver Province	158,687	159,692	+0.6
National Post[1]		158,250	
Montreal Gazette	159,108	156,379	-1.7
Toronto Sun	250,695	145,252	-42.1
Calgary Herald	121,690	130,595	+7.3
Winnipeg Free Press	145,214	127,305	-12.3
Ottawa Citizen	164,120	120,440	-26.6
Edmonton Journal	157,873	108,021	-31.6
Halifax Chronicle-Herald	96,744	106,395	+9.8
Hamilton Spectator	124,616	91,716	-26.4
Journal de Québec	103,047	90,874	-11.8
Le Soleil	102,920	80,720	-21.6
Windsor Star	84,206	59,849	-28.9
Edmonton Sun	82,145	46,201	-43.8

Source: Newspapers Canada (2010).

an ever-more prominent fact of the national experience and have helped to make the country a more vibrant and interesting place.

By comparison, newspapers have become a less prominent fact of the national experience. The emblem of the industry's vigour has always been circulation—the number of copies published each day—because this is the measure of the extent to which newspapers are implicated in the lives of

1 *The Financial Post*, which was absorbed by the *National Post* on the launch of the latter in 1998, had a circulation of 96,846 in 1995.

the citizenry, and the metric by which advertising rates are set. Consider the declines recorded in almost every Canadian newspaper market over the past fifteen-year span for which data are available.

Nationwide daily newspaper circulation in 1995 was 5.3 million copies (Dornan 1996, 64). In 2010, that figure had dropped to 3.8 million, a loss of 28 per cent (Newspapers Canada 2012). This, despite the fact that the population of Canada grew from 29.3 million in 1995 to 34.1 million in 2010, an increase of 4.8 million or more than 16 per cent.

Even the few instances of circulation growth in table 3.1 mask dispiriting facts for the industry. It is true that the *Calgary Herald*'s circulation grew by almost 9,000 between 1995 and 2010, but the circulation of the *Calgary Sun* dropped from 68,759 to 39,961—a loss of 28,798. So total newspaper circulation in Calgary dropped by some 20,000. And although the Halifax *Chronicle-Herald* showed a circulation gain of more than 9,000, in February 2008 its competitor, the Halifax *Daily News*, ceased publication. The *Daily News* had a circulation of some 20,000 and had lost money since Transcontinental acquired it in 2002. Total circulation in Halifax actually declined by more than 10,000.

While the circulation figures for the *Globe and Mail* have remained steady, even this is hardly good news for the industry as a whole. Most Canadian papers are parochial: they service a locality, an urban or regional area. A chief distinguishing feature of the *Globe and Mail* is that it aims for national reach. It features little of the local content that is typical of other dailies, such as classifieds, a real estate section, neighbourhood news, local sports, and so on. So the paper that has proven most resilient is the paper that bears least resemblance to the traditional model of the local daily.

The figures are even more striking when charted against the number of households in the country, as we can see from Table 3.2.

As table 3.2 demonstrates, it is not simply that daily circulation has been steadily dropping from a peak in the mid-1980s but that a cultural artifact, once a presence in every household, has become ever more marginalized. Extrapolated, table 3.2 charts a trajectory toward oblivion.

The Newspaper as Cultural and Industrial Form

The daily newspaper is a historical product of three symbiotic genres of content chasing after one target market: the public. Four hundred years ago, before there were newspapers as we know them today, there were printed accounts—"relations"—of notable occurrences of political and economic consequence, or of salacious interest. There were also pamphlets agitating heatedly for one policy or another, one political cause or another. And there were commercial sheets—advertisers—announcing nothing

Table 3.2 Total Daily Newspaper Paid Circulation and Total Households,
Canada, 1950–2010

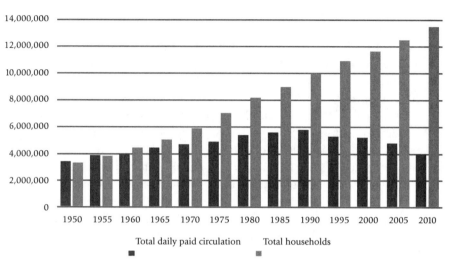

Total daily paid circulation   Total households

Source: Communic@tions Management Inc. (2011, 6).

more than goods for sale, services on offer, and investment opportunities.
The genius of the omnibus newspaper was to bring all three of these elements—news, editorial opinion, and advertising—together in one populist compendium so as to create an indispensable agency of public address.
In fact, the newspaper assists in the formation of "the public," a body
of citizens who can be addressed collectively and whose will or opinion
emerges first as a political and market force and then eventually as the
crucial determinant of democratic government.

The rising standing of the newspaper as means of social address is captured in the way in which it began to command revenue for its contents.
In England in the late 1700s, for example, the most popular and elaborate
entertainment of the day was the theatre. But in a city the size of London
how were audiences to know what plays were being staged, at which
theatres, and by which playwrights? At what time would the curtain rise?
What were the prices of admission? This information would be available
via newspapers, and so theatre patrons would have to pay to receive it, but
the newspaper publishers in turn paid the theatrical producers for the right
to publicize it. It was only in the early 1800s that the transaction became
reversed: as the daily paper became ascendant as a means of commanding
public attention, as it insinuated itself as an essential feature of public life,
it was the theatres (and indeed all commercial ventures) that depended on

the newspapers for publicity and so were required to pay publishers for the privilege of advertising their productions (Smith 1980, 11–12).

As it evolved, the newspaper proved particularly adept at creating editorial content that was simultaneously attractive to readers and a showcase and platform for advertisers. Over the course of the twentieth century, North American newspapers began to add subject-specific sections so as to capture more and more distinct advertising constituencies. The travel section, for example, fed on advertising from tour operators, air and cruise lines, hotel chains, even entire countries ("Come back to Jamaica"). Similarly, the automotive section, the real estate section, the fashion section, the health section, the food section, and so on supplied material of interest to readers—further inducements to purchase the paper—while at the same time attracting and providing a venue for an ever-expanding range of advertising clients. The picture of the world and one's community that the newspaper provided was very much a product of the imperatives of advertising. Content that could be married to advertising interests proliferated. Content that could not was anemic at best, or simply absent. So every daily newspaper in North America came to include a business section—filled with news for the investor, the manager, the executive, and the employer, and flush with advertising. But not one included a labour section that was addressed to the employee, was attentive to working conditions, and chronicled the experience of the unemployed. There was no commercial base, no advertising constituency, eager to underwrite such a section Monday to Friday, and hence it never existed.

At the same time, over the course of the twentieth century the daily paper accumulated more and more content to lure and occupy readers that had nothing to do with news: crossword puzzles, horoscopes, advice columns, bridge columns, stamp columns, comic strips, gardening and birdwatching columns. Usually these features were syndicated across newspapers throughout the continent and therefore cheap for any one newspaper to acquire.

The special preserve of newspapers was classified advertising, the small change of commercial transaction essential to the life of any community and extremely lucrative in sum. Even into the late 1990s, anyone looking to rent an apartment, sell a sofa, or post a death notice had almost no option but the local daily. Newspapers also carried display advertising for businesses, but classifieds were theirs alone. TV, radio, and magazines simply could not accommodate classified advertising.[2] To work properly, classifieds had to be

---

2 In the 1990s, however, the Quebec cable TV company Vidéotron attempted a classified ad channel in its offerings. Viewers sat through still photographs of used goods offered for sale by their owners,

readily at hand, continually updated, eminently searchable (hence the "classifications" in the columns of classifieds), and above all affordable.

By the end of the twentieth century, the paper-and-ink artifact that was delivered daily to one's doorstep was a vehicle for news and opinion, certainly, but the actual amount of news and opinion contained in its pages was dwarfed by advertising content, generic service journalism, and syndicated features.

It was also an offspring of the Industrial Revolution, which is why the newspaper's ascendance as an agency of public address occurred in the early 1800s. A mass circulation press was not possible without machines that could churn out thousands of copies a day. Steam power lent itself handily to powering any assembly line, from a saw mill to a printing press. The scale of manufacture also has to be such that the unit cost of production was next to negligible, and therefore the purchase cost of each copy was low, especially for an item that sought to resell itself to consumers every day, otherwise mass markets were impossible. So the newspaper, a product of an age in which even information had to be mechanically mass produced, was an object assembled and distributed by machines. It required a printing plant, regimes of factory production, fleets of distribution vehicles. The logic of production of the newspaper-as-artifact was identical to that of the ball-bearing industry. The difference was that ball bearings were not immediately out of date as soon as they were manufactured and did not presume to tell people what to think or what they should buy.

Also in common with the course of industry in the twentieth century, North American newspapers succumbed to chain and conglomerate ownership. In 2012, of the ninety-five daily newspapers published in Canada, exactly four remained independently owned: the Halifax *Chronicle-Herald*, *L'Acadie Nouvelle* of Caraquet, Montreal's *Le Devoir*, and the *Whitehorse Star*. The remaining ninety-one were owned by thirteen companies[3] that were either themselves newspaper groups or corporations with holdings beyond newspapers, or—as in the case of Torstar Corp.—both.[4]

Chain ownership and corporate control of newspapers became a prominent policy anxiety in the last quarter of the twentieth century, particularly given incidents in which the newspaper companies colluded to rationalize markets, and willful corporate proprietors were accused of skewing

---

everything from cars to furniture to wedding dresses. The channel had a hypnotic, voyeuristic appeal, but was ill-suited to actually vending goods because the content was not searchable.

3 Although one of these, FP Canadian Newspapers, owns only two titles: The *Winnipeg Free Press* and the *Brandon Sun*.

4 For a full and up-to-date list of newspaper ownership by company, see the Newspapers Canada website, http://www.newspaperscanada.ca.

the contents of their papers to advance their own political interests. Most famously, on August 27, 1980, the Southam chain's *Winnipeg Tribune* and the Thomson chain's *Ottawa Journal* simultaneously ceased publication, leaving Southam's *Ottawa Citizen* with a monopoly in the nation's capital and Thomson's *Free Press* with a monopoly in Winnipeg. The actions prompted the creation of the Royal Commission on Newspapers (the Kent Commission). In the 1990s and into the early 2000s both Conrad Black and the Asper family were controversially seen as exercising executive influence over the contents of their papers and firing journalists who did not share their points of view, actions that prompted the Standing Senate Committee on Transport and Communications to issue a report on the Canadian news media in 2006.

This policy concern and the investigations it spawned are a testament to how prominent newspapers once were. In the last quarter of the twentieth century, if one was in the market for municipal news, national news, political news, business news—anything, really—one all but had to turn to the local daily. And even the local paper was a poor vehicle for the direct expression of civic opinion from the grassroots. At most, it set aside a portion of the op-ed page every day—a fraction of a single page—for letters to the editor. A manifestly populist medium, therefore, nonetheless severely constrained and edited the ability of the populace to speak to one another via the pages of public address.

The persistent policy concern, then, about the newspaper industry was that the concentration of ownership and corporate control throttled the marketplace of information and opinion. Of course, concentration of ownership was a bad thing, said the report of the Special Senate Committee on the Mass Media, because it limited diversity of voices. "In a land of bubblegum forests and lollipop trees, every man would have his own newspaper or broadcasting station, devoted exclusively to programming that man's opinions and perceptions" (Senate of Canada 1970, 4). Such a land was inconceivable to the thoughtful senators of 1970.

The Circumstances of Decline

The newspaper's cultural place began to slip in the late 1950s, and the reason was plain: the advent of television. But while television ate into the time readers might spend with newspapers, it could not replace classified ads; it could not match the local and regional coverage offered by newspapers; and its news and documentary programming were proportionally minor genres in its overall output—in the main, commercial television was given over to entertainment.

As early as 1964, Marshall McLuhan had pointed out, "The classified ads (and stock market quotations) are the bedrock of the press. Should an alternative source of easy access to such diverse daily information be found, the press will fold" (McLuhan 1964, 207). That alternative arrived in 1995 when a San Jose software developer named Pierre Omidyar wrote a code he originally called AuctionWeb. On September 5, as a test, he announced on his home page that he wished to sell a broken laser pointer and invited bids. The item sold for $14.83. Omidyar had created the company that would become eBay: the first online auction house, a means to buy and sell precisely the sort of merchandise that heretofore could only be found at flea markets and in the newspaper classifieds.

Meanwhile, that same year, Craig Newmark, another software engineer in the San Francisco Bay area, began circulating an email list of upcoming social events to friends in his local community of programmers and Internet developers. As the number of postings and subscribers grew, people began to use the list to announce job openings in software and related fields, and to add more "categories" of listings (i.e., more "classifications"). In 1996, Newmark registered the domain name "craigslist.org" and the operation went live on the Web.

Such web-based advertising services are perfectly capable of catering to both local and global markets, simultaneously and effortlessly, in a way that newspapers simply could not. A purchaser in Helsinki can buy an item from a seller in Moose Jaw over eBay, while craigslist searches can be confined to particular cities. As well, even near-global web-based advertising operations can be run by limited staff, as opposed to the scores of advertising employees a single metropolitan newspaper was compelled to maintain. In 2012, even though operating in some 50 countries and logging some 30 billion page views per month, and with more than 200 million discussion forum postings under more than 100 topic headings, craigslist only employed about 30 staff members. More galling to the newspaper companies, craigslist began by charging users nothing to post to the service, and even today refuses to fully "monetize" itself, preferring as it says on its site to preserve "the relatively non-commercial nature, public service mission, and non-corporate culture of craigslist."

In 2005, a typical Canadian daily's operating revenue came in the form of local advertising at 34 per cent (these are display ads for businesses within the city the paper serves: furniture outlets, car dealerships, restaurants, etc.); national advertising at 20 per cent; classifieds at 17 per cent; circulation at 17 per cent; inserts and flyers at 6 per cent; and other operating revenue, such as custom printing, at 6 per cent (Bone 2007, 3).

Even if national advertising and circulation revenues were safe, which they were not, local and classified advertising accounted for 51 per cent of the traditional Canadian daily's revenue, and these were precisely the type of advertising most vulnerable to web-based services.

It is true that metropolitan newspapers remain at present a robust advertising vehicle—any medium that commands a daily circulation of 3.8 million is still a valuable means of reaching consumers. In 2010, Canadian daily newspapers recorded more than $3.1 billion in total revenues, an increase of 3.7 per cent over 2009. Of that total, $824 million came from circulation revenue; print advertising accounted for $2.1 billion; and advertising in online editions of papers brought in $213 million (Newspapers Canada 2012). However, the brute figures should be placed in a larger context of continuing trends, one of the most telling of which is the growth in digital advertising.

In 2010, for the first time, online advertising revenues in Canada surpassed daily newspapers' print advertising revenues: $2.23 billion to newspapers' $2.1 billion (Ernst and Young 2011, 5).

This is remarkable exponential growth. In 1997, total online advertising expenditure in Canada was a mere $10 million (Dornan 2003, 104). Thirteen years later, it amounted to $2.23 billion. But of this $2.23 billion, digital editions of daily newspapers captured only $213 million, or less than 10 per cent.

Something crucial and structural has changed in the newspaper industry. Together, eBay and its offshoots, along with craigslist, Kijiji (eBay's competitive answer to craigslist, launched in March 2005), and a multitude of other more specialized online merchandising forums, job markets, and professional networking venues such as LinkedIn (launched in May 2003), have smashed the newspaper industry's historical monopoly on classified advertising and destroyed the foundation on which the twentieth-century newspaper had been built. The advent of online advertising has permanently severed the union between editorial content and run-of-press advertising. There is no going back.

At the same time, the Web has brought about a proliferation of new and specialized sources of information. Why would the car enthusiast need to turn to the local daily's automotive section when there is a wealth of websites devoted to all aspects of the auto industry and with much richer content than a general interest newspaper might provide? What is the appeal of the local paper's movie section in a world of Rotten Tomatoes, Flixster, and IMDb, not to say an entire online industry of entertainment and celebrity gossip sites? Who needs the food section when one has Epicurious and

Table 3.3 Ten-Year Growth Trend of Online Advertising Revenue, 2001–2010

Year	2001	2002	2003	2004	2005	2006	2007	2008	2009	2010
Total Cdn. ($)Millions	86	176	237	364	562	900	1241	1602	1822	2232
% growth per year	-12	+105	+35	+54	+54	+60	+38	+29	+14	+23

Source: Ernst and Young (2011, 6).

myriad online cooking and recipe sites? In short, the Web has eclipsed the type of service journalism that the omnibus daily once used to marshal constituencies of advertising.

Meanwhile, the "land of bubblegum forests," once inconceivable to the Senate of Canada, in which "every man" could program "his own opinions and perceptions," has arrived. From Twitter to YouTube, Flickr to Tumblr to WordPress blogs, the contemporary moment heaves and seethes with self-expression. The Web has conferred powers of public address on any-one with a stake in political, social, cultural, and commercial affairs. If the newspaper was once the custodian of a cramped and limited form of public conversation, that conversation is now a cacophony with no need of a custodian.

Plus, in an age when information flashes to screens instantaneously, the industrial manufacture of physical copies of newspapers is ever more burdensome and wasteful, especially since no one reads a daily newspaper cover to cover. More importantly, the digital industries reproduce the model of broadcasting, in which the cost of the machinery of reception is passed on to the consumer, who is obliged to purchase the radio set, the television, the computer, or the tablet. In the print newspaper industry, the cost of putting the content into the hands of readers—the machinery of production and distribution—is borne by the producer.

Once the perfect amalgam of news, opinion, and advertising, in the space of a decade the newspaper's distinctive cultural attributes and market advantages have been stripped from it. It remains a prominent advertising vehicle, but it is no longer a *necessary* one, and its advertising capabilities are decidedly antiquated. It offers nothing much to advertis-ers beyond access to an undifferentiated public. An ad placed in the pages of a newspaper may or may not be seen by a reader of the paper. In the digital world, algorithms informed by the history of a reader's interests, information consumption patterns, and personal data (age, sex, etc.) can target commercial messages and consumption prompts to specific individuals. Hence the iTunes Genius function, book title recommendations from Chapters.ca, and Netflix viewing suggestions.

So, if the newspaper has been superseded as a forum for opinion and debate and compromised as an advertising vehicle, what does that leave it with? What does it have to offer that is not provided elsewhere with greater felicity, efficiency, and effect?

It leaves it with two things. First, news: an ongoing and, in the main, reliable chronicle of contemporary events from the local to the global. And, second, a physical distribution system. But even here the newspaper is saddled with difficulties.

The distribution system is an expense, although one that can still bring in revenue to the extent that newspapers are able to charge other parties to piggyback on it. A 2010 KubasPrimedia study found that "many newspaper executives ... indicated a strong desire to increase their pre-print advertising volume" (flyers and inserts) in the year ahead: "pre-print is probably two to three times more than online ad revenue. But the attention it gets is miniscule compared to digital" (Powell 2010). Executives can hardly be faulted for trying to find revenues wherever they can, but relying even further on an outmoded and hugely expensive physical distribution system to generate income is akin to making the most of the albatross around one's neck. Nor does the distribution system itself have anything to do with the remaining core product the newspaper industry has to sell: the news.

But the Web has also made possible the parasitic ventures known as aggregators that appropriate the content of newspapers even as they siphon readers away from the papers themselves, the most prominent being Google News. So, if one wants a daily digest of political news from media outlets across Canada, one can painstakingly survey the contents of papers from the Halifax *Chronicle-Herald* to the Victoria *Times Colonist* every morning, or one can simply visit nationalnewswatch.com, which generates no original content, culls the output of reporters and columnists from all the major Canadian dailies, and garners advertising tailored to a politically attentive readership.

At the same time, the editorial content most expensive for newspapers to produce is original reportage, while decades of downsizing to protect profit margins have compromised the industry's ability to generate it. As circulations continue to drop while advertising migrates to digital channels, newspapers do not have the wherewithal to reinvest in their newsrooms.

As well, the newspaper industry is its own victim. Driven by the imperial ambitions of corporate owners or a belief that the future of the media lay in

"technological convergence" (in which content providers and content distributors would benefit from symbiotic commingling), the 1990s in Canada
and elsewhere saw aggressive cross-media acquisitions. Hugely expensive,
these were financed by taking on massive debt, led to yet more editorial
downsizing (as reporters were expected to double up, feeding content to
both broadcast and print newsrooms), and in the end yielded little in the
way of profitable synergies. In both the United States and Canada, the debt
incurred by these purchases proved crushing and the legacy was bankruptcy.
The acquisition of the Southam chain by CanWest Global in 2000 led to the
insolvency of CanWest nine years later. The company that emerged in 2010
as the owner of the daily newspapers, Postmedia Network, remained saddled
with $700 million in debt. By September 2011 it had dismissed 14 per cent
of its workforce, 750 jobs, "bringing to 1,700 the total number of staff elim-
inated at the company since 2008" (McNish and Krashinsky 2011, n.p.).

The Digital Transition

The waning status of the newspaper has given rise to a new policy anxiety:
that democracy cannot sustain itself without the professional journalistic attention to contemporary affairs and civic institutions provided by
newspapers; that democracy requires an informed citizenry; and that if
newspapers fail, or persist only as diminished, enfeebled versions of what
they once were, then the governed will live in ignorance and woe to us all.
As American media scholars John Nichols and Robert McChesney put it:

> So this is where we stand: much of local and state
> government, whole federal departments and agencies,
> American activities around the world, the world itself—
> vast areas of great public concern—are either neglected
> or on the verge of neglect. Politicians and administrators
> will work increasingly without independent scrutiny and
> without public accountability. We are entering historical-
> ly uncharted territory in America, a country that from its
> founding has valued the press not merely as a watchdog
> but as the essential nurturer of an informed citizenry.
> The collapse of journalism and the democratic infrastruc-
> ture it sustains is not a development that anyone, except
> perhaps corrupt politicians and the interests they serve,
> looks forward to. (Nichols and McChesney 2009, n.p.)

There are those who reject the premises of the argument. When was the citizenry ever truly informed? The public has always preferred to be distracted than to be edified, and the media industries in North America have built their empires on that principle. What democracy demands is freedom of expression, not the persistence of a particular industrialized form of information provision. Why privilege an outmoded medium of cultural expression as integral to democratic practice over new and ascendant concourses of social communication, especially since no one congratulated the news media over the last half of the twentieth century for the exemplary job they were doing in covering the world, the nation, and the local community?[5]

Whether one is swayed by the arguments on either side, "concentration of ownership" no longer looms large as a policy concern with regard to the news media. It has been overshadowed by the fear that there will be nothing left to own, concentrated or not.

It is clear to the newspaper industry—it has been clear for some time—that it must make the transition from a medium printed on dead wood to digital delivery. But as yet this remains an unrealizable aspiration. The newspaper is squeezed in a historical moment, compelled to publish in both formats. It cannot yet abandon paper and ink, but at the same time it must project a robust digital presence, adopting and adapting to new formats and platforms of presentation. Rather than being able to rid itself of production expenses, the industry currently has to shoulder production expenses for two different media of presentation. Further, the industry cannot yet command nearly the advertising rates for its online incarnation that it has traditionally charged for its print editions: a 2012 Pew Research Center study found that US papers were losing seven dollars in print advertising for every one dollar they gained in digital revenue (Rosenstiel, Jurkowitz, and Ji 2012). Nor has the industry yet managed to settle on a model of compensation for its digital wares that is profitable for papers and acceptable to consumers. Should it cling to a subscription model, in which customers pay for blanket access to entire editions? Should it move to metered charges, selling access to individual articles the way iTunes sells individual songs? At present, the industry is monitoring the *New York Times*, which has instituted a "paywall"—allowing free access to a sample of the paper's contents but demanding compensation for the rest. (Even the phrase the industry uses—moving content "behind a paywall"—is revealing. It indicates a business with a siege mentality.) Part of the difficulty is that these payment models are being introduced to

---

5 For examples of arguments that contest the notion that newspapers are essential to democracy, see Eaves and Owen (2007) and Shafer (2009). For an account of twentieth-century academic discourse on the news media, see Dornan (1995).

a public that has become accustomed to "free" access to online content. A 2011 Canadian Media Research Consortium survey found that 85 per cent of Canadian Internet users consult digital news sources at least once a month, but 81 per cent of these said they would not pay if their preferred sources of online news began charging for access (CMRC 2011, 1–2). This is not to say that payment models will not eventually take hold—witness how television moved from "free" over-the-air delivery to payment to cable and satellite providers, to further pay-on-demand services and subscription fees through AppleTV, Netflix, BBC iPlayer, and others. But for the moment, neither digital advertising nor digital fees-for-access are producing revenues in the amounts necessary to maintain the news-gathering operations daily papers require if they are to have a news product to sell.

All of this supposes that the "newspaper" of tomorrow is a digital version—or reincarnation—of the Canadian print newspapers that exist today, produced by the same companies: Torstar, Quebecor/Sun Media, PostMedia, and the rest. But what is truly at issue is how and in what form journalism—the ongoing record of the real—will make the transition to a fully digitized future, not whether the Saskatoon *StarPhoenix* and its parent company will survive. MP3 players, file-sharing, Napster, iTunes, and the rest convulsed the music industry at the turn of the millennium, but they did not threaten songwriting. What they threatened were recording labels, A&R jobs, and the music industry, not the music itself.

In a world of laptops, tablets, and e-readers, the prohibitive costs associated with starting up a news outlet are all but gone. What one needs is the right idea: content that will attract readers willing to pay for it, or advertisers eager to reach those readers, or both, at rates sufficient to bankroll the expense of the staff to produce the content. Witness the rise of the *Huffington Post*, and its English- and French-language Canadian editions.

Launched in the US in 2005 as an aggregator and blog site, what the *Huffington Post* truly aggregated was eyeballs. It cost next to nothing to produce, since it fed off existing news operations and its original content took the form of opinion pieces written by individuals who were not paid, except in the publicity the platform afforded them. And yet within six years the publication was purchased by AOL for USD$315 million. Capitalizing on the Web's ability to cast both local and global nets, in 2008 the site began to launch city editions, beginning in Chicago and now ranging from San Francisco to Miami. In May 2011, it launched its first international edition in English Canada and added a French-language Quebec edition in February 2012. If nothing else, the *HuffPost* proves it is perfectly possible to establish a purely digital journalistic publication without the backing of a traditional media company.

Canada has seen its own such independent digital start-ups. Launched in 2011, *iPolitics* is a subscription-based service that covers federal political affairs with its own staff generating original content. *The Mark*, launched in May 2009, is a free outlet run by a small staff that commissions original content—though much of it unpaid. Again, neither of these publications attempts to reproduce the full range of offerings found in an omnibus daily: *iPolitics* limits itself to political news and commentary, while *The Mark* is an electronic version of a newspaper's op-ed page, collating opinion pieces on topical issues. But they vouchsafe a glimpse, at least, of online publications providing original versions of what were once different functions of the daily newspaper.

Is it possible for the offspring of the traditional daily to attract both subscribers and advertisers in the necessary numbers and at the rates required to cover basic costs? For example, can it

- pay for the cost of the editorial enterprise?
- make the publication sufficiently well-read within a community such that it provides a shared experience and a forum for citizens to speak to one another and to institutions of authority and governance?
- do so without the revenue that once derived from classified advertising?

Fortunately, there is a model of a publication that survives without classifieds and yet marries editorial content to advertising, that commands subscription revenue, and that both speaks to and helps to bring into being specific and distinct communities of interest. It is called the magazine.

Magazines

Though both magazines and newspapers are traditionally print media, predominantly vehicles for non-fiction (i.e., journalism), advertising supported and characterized by the regularity and frequency of their publication (in a way book titles are not), they are nonetheless very different types of periodicals. And as industries, they are quite distinct.

A daily publication like a newspaper produces a type of content that is ephemeral in the extreme. One can read a three-year-old copy of *Canadian Geographic* at the cottage and still derive value and pleasure from doing so. A three-day-old copy of a newspaper is just fire starter.

And although both are paper products, magazines are typically printed on much higher quality stock, most often glossy, with far superior colour, which makes them more durable, more attractive as objects, and more appealing as a medium for display advertising. As well, magazines are usually bound in some way—there is a spine to which the pages adhere, or the pages are centre-stapled, so that the publication is presented to the reader as a single, coherent artifact. The pages of a broadsheet are merely folded together, and different sections are detachable. This is not simply a cosmetic distinction. The omnibus newspaper is not a single publication at all but a compendium of different publications assembled under one title, all of which may address the same reader (everyman/woman) but in different guises of his or her life: sports fan, homemaker, taxpayer, moviegoer, parent, and so on. The magazine offers a thematically unified editorial product to constituencies of particular interest.

Crows and Finches

The daily newspaper has a citywide sales territory and considers everyone within it a potential customer. *Today's Bride* has a nationwide sales territory but targets a focused clientele of readers and advertisers. Or, to put it more succinctly, newspapers are general interest publications while magazines tend to be special interest publications.

Until the 1960s, the dominant magazines in North America were general interest titles. Like newspapers, they featured something for every member of every family. These were titles such as the *Saturday Evening Post*, *Collier's, Cosmopolitan, Look, Life*, and, in Canada, *Saturday Night, Weekend, Maclean's*. But when television captured advertising and the attention of audiences as a far superior form of the general interest magazine, the general interest titles either floundered and died or reinvented themselves (see Smith 1980, 135–141). So, in 1965, Helen Gurley Brown became the editor of *Cosmopolitan* and transformed the failing property into the Bible of the pink-collar single working woman, just as in Canada, in 1975, editor Peter C. Newman repositioned *Maclean's* from a monthly general interest magazine to a newsmagazine.

Magazines began to cater to as many specialized tastes and interests as might be imagined. The biggest selling titles became those whose specialized subject matter was of the broadest interest. So, for example, in the 1970s and 1980s one of the largest selling magazines in the US was *TV Guide*: in 1980 it had a circulation of 20 million. It was a specialized publication in that it was of interest only to those who owned a television, but every household owned a television.

Today, magazines published in Canada include *Brave Words and Bloody Knuckles* (for heavy metal aficionados), *Esprit de Corps* (about the Canadian military), *Canadian Horse Journal, Taps: The Beer Magazine, Canadian*

Lawyer, Celtic Kitchen, Comics and Gaming Monthly, Canadian Yachting, Canadian Coin News, Canada's Homeschool Guide, Adventure Kayak, and *Going Natural* (a quarterly for nudists).

Magazines are like Darwin's Galapagos finches: bewildering in their variety, continuously evolving, with new species springing up when circumstances allow, others adapting to the changing environment, and still others dying off when markets dry up or consumer interest wanes. Daily newspapers, by contrast, are like crows: they all look the same. In presentational form and categories of content, there is really no difference between the *Windsor Star* and the *Vancouver Sun*. And since the newspaper's habitat is the city—and since market economics are such that few cities can support more than two daily papers—the number of dailies in Canada is limited. Nor do dailies replenish themselves. They die, but they do not spring up. In 1911, Canada was home to 143 dailies. In 2012, there were 95. By comparison, there are so many magazines published in Canada, and of such variety, that it is difficult to enumerate them exactly. Magazines Canada tabulated that in 2009 there were 1,276 magazine titles published in Canada, 76 per cent of them in English and 24 per cent in French, an increase of 36 per cent over 1999 (Magazines Canada 2011a, 6). Elsewhere, the organization has claimed, "There are 2,053 Canadian magazine titles" (Magazines Canada 2011c, n.p.).

Table 3.4 Number of Print Magazine Launches in Canada vs. Number of Closings

Year	Launches	Closings
2011	19	9
2010	13	2
2009	17	37
2008	74	29
2007	75	28
2006	82	24
2005	93	33
2004	139	34
2003	101	47
2002	83	50
2001	69	43
2000	81	60

Source: Mastheadonline.com (2012).

US Competition

While competition from US newspapers has never been a significant factor for the newspaper industry, US competition has long been the predominant concern for magazines. The priority of cultural policy has been to provide the industry with support so as to survive in the face of the availability and attractiveness of the American product. Nonetheless, US titles continue to swamp newsstands in Canada and dominate single-issue sales, "taking approximately 95 per cent of sales and shelf presence" (Magazines Canada 2008, 3). Canadian titles, by comparison, dominate subscription sales in Canada.

Support for the magazine industry has taken the form of tariffs on American competitors, taxation policy and subsidies—or a combination thereof. The government cannot prohibit the sale of US magazines, nor would it wish to. Imagine a Canada in which it was not possible to buy *Scientific American, The New Yorker, Sports Illustrated, The Utne Reader, Teen Beat, Ms., MAD.* It would be an alien and culturally impoverished place. It would be Cuba. The fear has not been the availability of American magazines but the prospect of what are called "split-run" editions of US titles selling in Canada.

A split-run is a "Canadianized" version of a US title—essentially the American magazine with some Canadian content added—that also solicits Canadian advertising. What would happen to the market for *Canadian Geographic* among readers and advertisers if there were a "Canadian" version of the much more muscular *National Geographic*? In 1965 the Canadian government enacted Custom Tariff 9958, explicitly to prevent split-run editions of foreign publications from being sold in Canada (Dubinsky 1996, 43).

In 1975, the government passed Bill C-58, an amendment to the *Income Tax Act*, disallowing tax deductions for advertising expenditures in any but Canadian-owned publications and broadcasting outlets. This meant that a business in Toronto, for example, was perfectly at liberty to advertise on a television station in Buffalo, NY, with viewership in Toronto but would not be able to deduct the cost of doing so as a business expense. The amendment had the effect of repatriating an estimated $20 million a year that had been flowing out of the country, funnelling that money to Canadian broadcasters and to newspaper and magazine publishers.

In 1993, however, *Sports Illustrated* contracted to print and distribute a Canadianized split-run edition in Canada. Because the edition was being run off Canadian printing presses, no copies of the magazine were crossing the border, and so Custom Tariff 9958 did not apply. And although it was not a deductible expense for Canadian businesses to advertise in it, the rates offered were so cheap—since the bulk of the magazine had already been produced for the US market, and so its costs had been already more than recouped—that the *Income Tax Act* disincentives were ineffectual.

The Canadian government responded by imposing punitive taxes on the sale of any publication that did not contain at least 80 per cent Canadian content. But since this meant only foreign publications were affected, it triggered a full-blown trade dispute, with the United States threatening retaliatory measures against Canadian exports such as textiles and wood products. The United States took the dispute to the World Trade Organization, which ruled in 1997 that the tax amounted to unfair restraint of trade.

In response, the Canadian authorities instituted a number of measures designed to comply with the WTO ruling while at the same time providing support to the Canadian magazine industry. Canadian businesses were free to advertise in both domestic and foreign magazines, though they could claim only 50 per cent of the expense of doing so if the ads were placed in periodicals with less than 80 per cent original Canadian content. At the same time, in 2000, the government introduced direct financial subsidies to Canadian magazines through the Canadian Magazine Fund.[6]

An element of this supportive regime included postal subsidies, through what was called the Publications Assistance Program. Because most magazines cater to communities of interest rather than, like newspapers, to geographic areas, the members of these various communities may be spread across the country. The readers of *Our Times* or *Opera Canada* are not confined to one city; Canada is a vast country, and most Canadian magazines are sold by subscription rather than from newsstands. Postal subsidies have long been an important measure in the effort to put domestic magazines in the hands of Canadian readers. However, Canada Post was increasingly chafed at having to participate in the subsidy regime and, in 2006, announced its intention to end its $15 million annual contribution. The Conservative government of Stephen Harper subsequently phased out the Canada Magazine Fund and the Publications Assistance Program and replaced it in 2010 with the Canada Periodical Fund (Canadian Heritage 2009).

The Periodical Fund allocates $72 million per year for assistance to individual magazines, along with a further $1.5 million for business innovation and $2 million for initiatives to benefit the industry as a

---

6 For a thorough examination of the design and implementation of the Canadian Magazine Fund and its effect over the ten years of its existence, see McCann (2011).

whole. There are three prominent features of the new regime: First, it rewards success in the marketplace, disbursing funds so that the best-selling magazines receive proportionally more money. Second, magazines with circulations under 5,000 are no longer eligible for funding, meaning that small literary journals must now appeal to bodies such as the Ontario Arts Council for support. Third, postal subsidies are no longer part of the regime, but monies can be used to develop means of digital presentation and distribution.

The Digital Transition

The number of different Canadian magazines currently available might be growing but the sales of individual titles are experiencing the same order of decline as newspapers, with the two industries now caught in the same dilemma. The paper-and-ink product is expensive to produce and distribute, but readers are not yet ready for a purely electronic presentation of content. This is probably truer for magazines than newspapers. Subscribers genuinely like their magazines and have a loyalty to them that few newspapers inspire. One may read a city daily because one has to in order to stay abreast of local affairs, but no one *has* to read *The Antigonish Review* or *Inside Motorcycles*. People read such magazines because they want to. And most magazines, aesthetically, are more handsome than newspapers, with a greater sense of design, which makes their readers appreciate them as physical objects. So it may be more difficult for magazines to effect a transition to digital delivery.

As table 3.5 demonstrates, these are steep and worrying declines over a span of only four years. A longer view makes them all the more stark: ten years ago the circulation of *Reader's Digest* was more than a million. Under the tenure of legendary editor Doris Anderson (1957–1977), *Chatelaine* also had a circulation of more than a million.

But the objection to digital delivery and presentation of magazines is surely a matter of ingrained habit (which can be overcome). Plus, until very recently the online presentation of newspapers and magazines has been via websites viewed on desktop and laptop computers. In the main, the aesthetics of such digital publications have been cluttered, ungainly, and unattractive.

Until now. In April 2010, Apple launched the iPad, followed by a second generation of the machine in March 2011, and a third, with an even higher resolution screen, in March 2012. The Kindle Fire, with a colour screen, debuted in September 2011. Samsung, Motorola, and Asus all introduced tablets, running on the Android operating system. The experience of reading a magazine—or a newspaper—on one of these mobile

Table 3.5 Circulation for the Top 5 Canadian Magazine Titles by Market (in thousands): 2007–2011

English	2007	2008	2009	2010	2011	% change
Reader's Digest	868	851	807	663	593	-31
Chatelaine	576	552	522	507	536	-7
Canadian Living	538	521	516	508	508	-5.6
Maclean's	356	347	359	349	330	-7.3
Canadian House & Home	244	234	226	240	242	-0.8
French						
Sèlection	234	232	220	186	171	-26.9
Coup de Pouce	228	227	212	211	205	-10
Châtelaine	201	198	192	175	173	-13.9
L'actualité	178	173	169	156	163	-8.4
Bel âge	139	140	134	125	125	-10

Source: Audit Bureau of Circulations

devices is in many ways superior to the experience of reading the printed product. *The New Yorker* app designed for the iPad by Adobe is a gorgeous thing, an exquisite recreation of the print version that also takes advantage of the tablet's capabilities to add touch-screen features such as video, annotated slideshows, and audio. Advertisers can only exult in the visual quality of the ads, as well as touch-through capabilities to route interested readers to more detailed information on products. The format allows the reader to comment, like, dislike, post, Tweet, flip-forward, and otherwise churn the content in ways impossible with sheets of paper, no matter how glossy. And, at present, access to the app comes with one's subscription to the print magazine, so as to ease and usher the transition to digital.

Just as once upon a time people bought newspapers because they carried crossword puzzles, today people pay a small charge to play Angry Birds on their mobile devices. By the same token, the tablets are show windows for magazines, newspapers, and new titles: they are a means of both promoting the product and accessing it. As of April 5, 2012, there were 2,226 magazine titles available via Apple's Newsstand. By June 15, 2012, there were 3,302.

In Canada in 2011, less than two years after the debut of the iPad, 9 per cent of adults owned tablets, and a Magazines Canada survey found that 33 per cent of those who didn't professed their intention to buy one within the next six months (Magazines Canada 2011b, 7). The Pew Centre found that 30 per cent of tablet owners in the US spent *more* time getting news than before they owned the device (Pew Research Centre 2011).

On March 9, 2012, the *Toronto Tempest* made its debut: a tablet-only title available through the App Store and Google Marketplace. "Until the tablet," said co-editor James Burrows, "the costs of launching an independent magazine were prohibitive. We expect to be the first of many new, young, and interesting magazines that take advantage of this space" (Mastheadonline 2012). Meanwhile, in November 2011, *National Geographic* topped more than 100,000 paid digital subscribers (Kinsman 2011). *Cosmopolitan* did so in March 2012, and did so by charging more for the digital edition than the print magazine and charging for it separately rather than offering free digital access with the print subscription (Roy 2012).

The Kairotic Moment or Reasons for Optimism

The ancient Greeks bequeathed us two concepts of time: *chronos* and *kairos*. The first refers to the more familiar chronological flow of time, but the second refers to a moment of temporal suspension, an interlude in which something special occurs, "a passing instant when an opening appears which must be driven through with force if success is to be achieved" (White 1987, 13).

Given the inexorable trajectory of decline for traditional newspapers and magazines, given the inflection point that has been reached with digital advertising expenditures surpassing those of print, and given the exponential growth of tablet sales, it would appear that newspapers and magazines have arrived at their kairotic moment. Individual titles may not survive, entire companies may disappear, and what emerges on the other side may bear only passing resemblance to the printed periodicals of the twentieth century, but finally there is reason for optimism if not indeed exuberance about the prospects for newspapers and magazines as cultural industries.

References

Bone, Allison. 2007. "Revenue Fluctuations for Newspaper Publishers." Analytical Paper Series – Service Industries Division. Ottawa: Statistics Canada.

Canadian Heritage. 2009. "The Government of Canada Creates Canada Periodical Fund to Better Support Magazines and Community Newspapers." Accessed February 17, 2012. http://www.pch.gc.ca/eng/1294862436162/1294862436165

Canadian Magazines. 2012. "Torstar Buys a Piece of Blue Ant, which Bought a Piece of Quarto, and So On" (January 2). http://canadianmags.blogspot.ca/2012/01/torstar-buys-piece-of-blue-ant-which.html xxx

Canadian Media Research Consortium (CMRC). 2011. "Canadian Consumers Unwilling to Pay for

News Online" http://www.cmrcccrm.ca/en/projects/PayingForTheNews.htm (March 29).

Communic@tions Management. 2011. "Sixty Years of Daily Newspaper Circulation Trends: Canada, United States, United Kingdom" (May 6). http://media-cmi.com/

Craigslist. 2012a. "A History of Craigslist." Accessed March 24, 2012. http://www.craigslistme.net/article/22/History_of_Craigslist.html

———. 2012b. "About Fact sheet." Accessed March 24, 2012. http://www.craigslist.org/about/factsheet

Denley, Randall. 2012. "Did You Know Big Buck Magazine Is Getting Your Bucks?" *Ottawa Citizen*, January 10. http://www2.canada.com/ottawacitizen/news/archives/story.html?id=fc6d5bac-c691-4d44-888b-c620f1ca5852

Dornan, Christopher. 1995. "Sounding the Alarm." *Media Studies Journal* Vol. 9, No. 3.: 181–194.

———. 1996. "Newspaper Publishing." In *The Cultural Industries in Canada: Problems, Policies and Prospects*, ed. Michael Dorland, 60–92. Toronto: James Lorimer & Company Publishers.

———. 2003. "Printed Matter." In *How Canadians Communicate*, ed. David Taras, Frits Pannekoek, and maria Bakardjieva, 97–120. Calgary: University of Calgary Press.

———. 2007. "Other People's Money: The Debate Over Foreign Ownership in the Media." In *How Canadians Communicate II*, ed. David Taras, Maria Bakardjieva, and Frits Pannekoek. Calgary: University of Calgary Press.

Dubinsky, Lon. 1996. "Periodical Publishing." In *The Cultural Industries in Canada: Problems, Policies and Prospects*, ed. Michael Dorland. Toronto: James Lorimer & Company Publishers.

Eaves, David, and Taylor Owen. 2007. "Missing the Link: Why Old Media Still Doesn't Get the Internet." http://missingthelink.net/missing-the-link-complete-version/

Ernst and Young. 2011. "2010 Actual and 2011 Estimated Canadian Online Advertising Revenue Survey." Sponsored by the Interactive Advertising Bureau of Canada.

Gillin, Paul. http://newspaperdeathwatch.com/

Holahan, Catherine. 2007. "eBay takes on Craigslist." *Bloomberg Businessweek*, July 6. http://www.businessweek.com/technology/content/jul2007/tc2007075_395980.htm

Hsiao, Aron. "How did eBay Start?" *About.com* http://ebay.about.com/od/ebaylifestyle/a/el_history.htm Retrieved March 24, 2012.

Kinsman, Matt. 2011. "National Geographic Magazine Exceeds 100,000 Paid Digital Subscribers." *Folio*, November 1. http://www.foliomag.com/2011/national-geographic-magazine-exceeds-100-000-paid-digital-subscribers

Magazines Canada. 2008. "Magazines Canada Submission to Canadian Heritage Magazine Policy Review" (April). https://www.magazinescanada.ca/public_affaires_cpf

———. 2010. "Magazines Canada Submission to House of Commons Standing Committee on Finance; Pre-Budget Consultations" (August).

———. 2011a. "Consumer Magazine Factbook 2011." http://www.magazinescanada.ca/advertising/fact_books

———. 2011b. "Digital Magazine Factbook 2011." http://www.magazinescanada.ca/advertising/fact_books

———. 2011c. "Canada's Magazine Policy is Working." http://www.magazinescanada.ca/public_affairs/news?news_id=1007

Mastheadonline.com. 2011. "Homemakers Magazine Coming to an End" (October 27). http://www.mastheadonline.com/news/2011/20111027666.shtml

———. 2012a. "Masthead Tally: Launches increase 46% in 2011" (March 2). www.mastheadonline.com/news/2012/20120301903.shtml

———. 2012b. "'Canada's first tablet-only politics and culture magazine' launches this Friday. March 7. http://www.mastheadonline.com/news/2012/20120307775.shtml

McCann, Julie. 2011. "The Canada Magazine Fund: An 'Alien' Support Method in a Challenged Industry." Master of Journalism thesis, Carleton University.

McGinn, Dave. 2009. "The gospel according to The Mark." *The Daily Planet.com*. May 25. http://www.thedailyplanet.com/index.php?option=com_content&view=article&id=252:the-gospel-according-to-the-mark&catid=52:industry-links&Itemid=282

McLuhan, Marshall. 1964. *Understanding Media: The Extensions of Man*. New York: McGraw Hill.

McNish, Jaquie, and Susan Krashinsky. 2011. "The Glitch in Postmedia's Digital Switch." *Globe and Mail*, September 29. http://www.theglobeandmail.com/report-on-business/the-glitch-in-postmedias-digital-switch/article2185559/

Mutter, Alan D. 2012. "Publishers Are Flubbing the iPad." *Editor and Publisher*, February 7. http://

www.editorandpublisher.com/Newsletter/Article/Publishers-Are-Flubbing-The-iPad

Newspapers Canada. 2012. www.newspaperscanada.ca

Nichols, John, and Robert W. McChesney. 2009. "The Death and Life of Great American Newspapers." *The Nation*, April 6. http://www.thenation.com/article/death-and-life-great-american-newspapers

Pew Research Centre. Project for Excellence in Journalism. 2011. "The Tablet Revolution." (October 25). http://www.journalism.org/analysis_report/tablet

Quebecor. "The Largest Magazine Publisher in Québec." http://www.quebecor.com/en/content/largest-magazine-publisher-qu%C3%A9bec

Powell, Chris. 2010. "KubasPrimedia Study Shows Newspaper Industry Optimistic about 2011." *MarketingMag.ca*, December 1. http://www.marketingmag.ca/news/media-news/kubasprimedia-study-shows-newspaper-industry-optimistic-about-2011-6252

Rosenstiel, Tom, Mark Jurkowitz, and Hong Ji. 2012. "The Search for a New Business Model." Pew Research Centre, Project for Excellence in Journalism (March 5). http://www.journalism.org/analysis_report/search_new_business_model

Roy, Jessica. 2012. "How Cosmopolitan Netted 100,000 Paid Digital Subscriptions." *10,000 Words*. http://www.mediabistro.com/10000words/how-cosmopolitan-netted-100000-paid-digital-subscriptions_b11422

Shafer, Jack. 2009. "Democracy's Cheat Sheet? It's Time to Kill the Idea that Newspapers Are Essential to Democracy." *Slate*, March 27. http://www.slate.com/articles/news_and_politics/press_box/2009/03/democracys_cheat_sheet.html

Senate of Canada. 1970. *Uncertain Mirror: Report of the Special Senate Committee on the Mass Media.* Vol. 1. Ottawa: Queen's Printer.

Smith, Anthony. 1980. *Goodbye Gutenberg: The Newspaper Revolution of the 1980s.* Oxford: Oxford University Press.

Toronto Star/Associated Press. 2009. "Reader's Digest to file for bankruptcy." August 17. http://www.thestar.com/business/article/682242

Transcontinental. "A key part of your daily life … as a publisher." http://tctranscontinental.com/en/2-product/2-2-1-magazines.html

White, E.C. 1987. *Kaironomia: On the Will to Invent.* Ithaca, NY: Cornell University Press.

4

Video Games Production: Level Up

Greig de Peuter

Level Up

In the early 1980s, a pair of enterprising suburban Vancouver teens, backed by a $4,000 investment, made one of Canada's first homegrown commercial computer games—*Evolution*. In 1991 the company the duo founded, Distinctive Software, was sold to California's Electronic Arts, one of the biggest game companies on the planet. The purchase foreshadowed the role of foreign-owned firms in Canada's video game industry in years to come. Leaping to the early 2000s, several best-selling games—*Assassin's Creed, Neverwinter Nights, NHL, Splinter Cell*—were made in Canada. By 2010, amid a generally gloomy economic outlook, the nation's game sector was grabbing headlines: Canada surpassed Britain as the world's third-largest game-industry employer after the US and Japan. Canada has, in game-speak, "levelled up." It is a significant location in a digital play business whose global revenue is anticipated to hit a record-busting $82 billion in 2015 (Cross 2011, 3).

This chapter is a snapshot of this relative newcomer to the cultural industries in Canada. It begins by surveying the industrial structure, gaming platforms, and player demographics characterizing the game business. The chapter goes on to sketch out video game production in Canada through a discussion touching on three themes: *players*, which identifies some of the key corporate actors and state agencies involved in game development in Canada; *producers*, which zooms in on people making games for a living; and *places*, which highlights the agglomeration of game industry activity in a handful of Canadian cities. The players, producers, and places animating Canada's video game industry exemplify the logic— and the tensions—of the so-called creative economy.

Shorthand for an ascendant policy-oriented discourse, the "creative economy" can be roughly anatomized as "creative industries," "creative class," and "creative city." Creative industries is a vast canopy covering, among other sectors, film, music, fashion, architecture, design, and software—like video games. The creative-industries concept initially took root among governments anxious to counteract the economic effects of national manufacturing industries in decline in the wake of globalization. Official interest in creative industries turns on their market expediency, namely, to generate intellectual property, spur entrepreneurial behaviour, and fuel job creation. This leads into the "creative class" (Florida 2003) of engineers, architects, writers, artists, and scientists depicted as passionately devoted to cerebrally rewarding, highly skilled, and comfortably paid jobs—like developing video games. The creative class, this line of thinking goes, impels a "creative city" (Landry 2008) offering amenities attractive to the "talent" that powers the lucrative creative industries, industries that promise to "revitalize" neighbourhoods—like those where game studios set up. It is in the political economy of creativity that the real-world stakes of virtual games are won or lost.

Play Factory

Digital games were hacked into being a half-century ago in Cold War America. Tinkering computer programmers and university students working in defence facilities and academic laboratories transformed bulky military mainframes and their rudimentary electronic screens into playthings. Part of a then emerging occupational group of high-tech experts, these pioneers adapted machinery intended for war and work into a medium for leisure. It was a masterstroke of innovation. Later in the 1970s, legendary companies like Atari converted the interactive game into a commodity. Over the next two decades, Nintendo, Sega, and Sony became household names as they and other firms engineered the hardware, developed the software, and popularized the pastime of video game play.

Encompassing digital play on a dedicated console, personal computer, or mobile device, video games currently comprise "the fastest growing entertainment industry across the globe" (Afan 2010, 45). The sector's annual sales surpassed the $50 billion mark in 2010 (Cross 2011, 3). Occupied by some of the world's largest multinationals, like Microsoft and Sony, gaming's blockbuster scale was confirmed in 2011 when the shooter game *Call of Duty: Modern Warfare 3* rang up over USD$1 billion in sales within a trim sixteen days—beating out the film *Avatar*, which made its first billion in seventeen days (Takahashi 2011). Revenue-wise, games not only rival the Hollywood box office but leapfrog it to truly otherworldly realms of profit-making: at its peak, twelve million subscribers populated *World of Warcraft*, one of the massively multiplayer online games, or MMOs, whose systems for exchanging in-game virtual goods for increasing quantities of real-world cash represent the latest area of inquiry for twenty-first-century economists (Castronova 2005).

Video games are played on a widening range of platforms. The biggest moneymaker is the console. A trio of firms traditionally dominates this wing of the business: today, Sony with its PlayStation 3, Microsoft with its Xbox 360, and Nintendo with its Wii. While primarily running store-bought game discs, these Internet-linked machines offer increasing online content downloadable from network services like Xbox Live. Console manufacturers practise perpetual upgrade economics by soliciting gamers every five years or so to replace their ostensibly dated device with a "next-generation" console. Smaller in market terms, but nonetheless significant, are games played on PCs, particularly those providing entry to MMOs. A mainstay for digital play, especially among younger gamers, is the hand-held console, such as the Sony Vita. Currently, however, social networks and mobile phones are the highest-growth platforms. This multiplication of game media has made the contemporary video game industry and its products increasingly variegated and ubiquitous—much like gamers themselves.

The rise of the "casual" gamer should lay to rest any lingering assumption that gaming's influence is restricted to a narrow demographic of pre-adolescent males. According to a Canadian study by the NPD Group for the Entertainment Software Association of Canada, nearly 60 per cent of those polled had played a video game in the past month (ESAC 2011, 13). Many children receive their introduction to new media culture via gaming, with about 80 per cent of youth aged six to seventeen having reported playing a game in the last four weeks (*ibid.*). The average age of the gamer is, according to the same report, thirty-three (*ibid.*). Those who grow up gaming continue to do so as adults, albeit in smaller doses, which is not surprising given the number of hours required to learn—let alone complete—many console games. The medium's long-standing testosterone bias has not vanished: 38 per cent of gamers in Canada are female, 62 per cent are male (*ibid.*, 14). The long-term trend is, however, an increase in the gender parity of the gaming populace.

As the gaming market grows, so too does interest in it from other cultural industries. Music-based games like *Rock Band* and game-based movies like *Resident Evil* indicate that games are not just competing with other media but are melding with them in a convergent entertainment complex. The abundance of individual games circulating through this complex defies generalization...up to a point. Military-themed games, sport titles, licensed properties, and sequels to proven hits account for the lion's share of game output. Interest is broadening, however, around "serious games," a genre lumping together games designed to serve a hodgepodge of government, corporate, and activist objectives, from fighter-jet training to personal-finance management to climate-change awareness. Early on, game culture was the target of moral panic over social isolation and copycat crime. Today, gaming is more likely to be embraced by a wider demographic. The U-turn is striking in accounts by business gurus arguing that gaming is training for the contemporary workplace (Beck and Wade 2004) and by civic-minded designers proposing that game space is a staging ground for confronting a variety of social ills (McGonigal 2011).

The game sector is not a homogeneous entity. Its industrial structure can be schematically broken down into five areas. First is console manufacturing. The big brands that make up the console oligopoly create proprietary hardware that is typically sold at a loss in order to build consumer market share and extract profits from system software receipts. Second is game production. This is a more variegated field. It includes the console makers that develop games at in-house studios for their own platforms as well as third-party studios that produce games on a contractual basis for game publishers (akin to a music artist and a record label). Game production is increasingly concentrated—a handful of publishers regularly dominate the bestseller list; expensive—budgets for AAA console titles can easily exceed $50 million; and risky—a rule of thumb is that 10 per cent of the games earn 90 per cent of the money. Third is middleware, which includes the making and licensing of software tools needed to develop a game. Fourth are ancillary services such as motion-capture, sound recording, and game-testing. Fifth is distribution, both to retail stores and via online portals like app stores. The latter are offering small studios fresh opportunities to distribute their products independently of large publishers.

The play factory is a transnational operation. Historically, game sales have been highest in North America, Western Europe, and Japan. Gaming's expansion into Asia is, however, giving the market a new territorial dimension. South Korea is one of the most intensive online gaming cultures in the world. And while the US remains the largest national market, China is now in second slot, trailed by Japan (Cross 2011, 4). Vast new player populations are opening up, but for the majority of the earth's inhabitants, the latest console is out of reach. Large-scale pirating and the market in old consoles nevertheless give games a circulation beyond the world's most affluent regions, into Latin America, the Middle East, and Southern Asia.

It is not just the consumption of games but their production that is globalizing. Consoles, hand-helds, and mobile devices are assembled in factories in Southern China, Eastern Europe, and Latin America. A new game might be conceived of in one country, funded from another, developed with inputs from multiple far-flung locales, and marketed to a global audience. And, ultimately, game media rely on natural resources extracted from sources such as African mines, and wind up dumped in one of the e-waste villages dotting the Global South. So, although the play factory is spread around the planet, its pleasures and its pains are very unevenly distributed.

Players

Canada is a major node in the global games industry. National consumer spending on game media neared $2 billion in 2010 (ESAC 2011, 16). It is, however, Canada's status as a premiere location for game *production* that garners most attention in creative-economy discourse. Canadian game businesses have an estimated annual direct economic impact of $1.7 billion (*ibid.*, 4). From development to publishing, middleware, ancillary services, and distribution, there are some 350 game companies in Canada (SECOR 2011, 6–8). The majority—nearly 75 per cent—are small enterprises with a staff of no more than twenty. Next is a thin band of medium-sized firms—about 15 per cent of the total—averaging around seventy employees. The remaining 10 per cent or so are large companies with an average of 220 staff. This group includes a small number of disproportionately large game firms, including two of the world's biggest: EA and Ubisoft, billion-dollar businesses with approximately 2,500 employees a piece in Canada. The Canadian industry is, then, lopsided: there's a vast pool of companies, but a fraction of them commands the bulk of revenues and employees.

While this establishes the industry's scope, grasping its political-economic dimensions requires a different set of emphases, including publisher power, foreign finance, collective capital, and state subsidy.

Discourse on creative industries emphasizes the importance of small- and medium-sized enterprises (SMEs) as sources of innovation and employment. Canada's game development sector is home to many renowned SMEs, and micro-sized operations specializing in emerging platforms are proliferating. Yet, in terms of economic impact, the industry is dominated by the production of console games. In this domain it is massive conglomerates rather than SMEs that are the decisive players. Here what must be highlighted is the nature of the relationship between developers and publishers. Developers generally create games under contract to publishers. Publishers include the console makers—Microsoft, Nintendo, Sony—as well as third-party companies—such as Capcom, EA, THQ, and Take-Two—that usually operate in-house studios and also hire outside studios on a project-by-project basis to make games for their label.

The power relationship between developers and publishers is uneven. Publishers finance game development, exercise control over the decisions about which games get made, and market the end product. This tilted arrangement is often justified on the grounds that console game creation is pricey, and risky. It can take a team of fifty or more developers between one and three years to produce a title on a budget between $20 and $100 million. And most games sink without a trace. Typically, a developer receives incremental payments from a publisher as development milestones are reached. Economically dependent in this way, developers can be quite precarious, as they are beholden to publishers that can technically pull the plug on a development project at any time. The locus of this publisher power resides outside Canada.

Creative industries are promoted under the banner of national competitiveness—yet the Canadian game industry illustrates the centrality of international corporate players. The country's game sector is overwhelmingly export-oriented. More than half of companies report that 90–100 per cent of their revenue derives from clients outside Canada, particularly the US (Hickling Arthurs Low 2009, 6). Most game firms in Canada—85 per cent—are Canadian-owned (Gouglas et al. 2010, 6). But with notable exceptions such as Behaviour Interactive (Quebec) and Silicon Knights (Ontario), Canadian-owned studios are small-scale entities (*ibid.*). At the medium to large scale, Canada's game companies are predominately foreign-owned. Several of the world's biggest developer-publisher conglomerates—such as Activision Blizzard (France), Capcom (Japan), Eidos (UK), and Take-Two (US)—have Canadian outposts.

Foreign game capital has deepened its presence in Canada by acquiring homegrown developers. This pattern was set early on when California's EA purchased BC's fledgling Distinctive Software in 1991. This was the crucible of EA Canada, which today has a sprawling studio in the Vancouver suburb of Burnaby where it churns out cash-cow sport games exclusively for its US parent. In the past decade or so, EA has gobbled up several Canadian studios with proven commercial hits, including Edmonton's BioWare—regarded, somewhat inaccurately, as "the crown jewel of

Canada's video game industry" (Nowak 2011a, n.p.). Concentration of ownership in foreign hands is a trend accelerating across the sector. While EA is at its forefront, it is not alone: Disney (US), Nexon (South Korea), Ubisoft (France), Vivendi SA (France), among other mega-publishers, have all bought out junior Canadian studios. On the upside, the acquired studios "gain access to their parents' deeper pockets," but this comes at the cost of "a degree of control" when the seat of decision-making power shifts (Hon 2009).

Canadian studios frequently create titles based on licensed property whose ownership lies outside Canada. While developing a game based on an original idea is a dream of every game designer, the funds necessary to realize it are difficult to access. Given the profit potential of a hit game based on new IP, however, finance capital is beginning to circle the game sector. Led by former EA executives, Vanedge Capital is an investment fund that steers capital, or money in search of money, to studios working on promising projects (Kyllo 2009a). Headquartered in BC with an office in Shanghai, Vanedge taps global investors capable of putting in a minimum of CAD $2 million. Its model is straightforward: finance game firms "with the expectation that we're going to sell them" (and hence be absorbed by conglomerates) "or have them go public" (and hence be listed on a stock exchange open to international participation) (cited in *ibid.*, n.p.). Vanedge illustrates the imprecision of "foreign finance" as an identifier of the geographic source, and beneficiaries, of game capital in Canada. It is a truly transnational playing field.

Game workers have not formed a labour union to collectively protect their interests. In contrast, collective game capital—or game companies as a group—pursues its interests through an industry association. The Entertainment Software Association of Canada (ESAC) is made up of eleven companies, a lean membership that nonetheless accounts for 90 per cent of game media sales in the country (ESAC 2011). "Educating" policy-makers is one of ESAC's goals ("Canada World's Third Largest..." 2011). It sponsors and publicizes reports on the industry's contribution to the Canadian economy, which spotlight gaming's high rate of growth as compared to other sectors (SECOR 2011, 6). While ESAC has multiple policy objectives, from tightening anti-piracy legislation to relaxing trade barriers, collective game capital's meta-policy position is perhaps most frankly expressed in this statement from a 2008 ESAC-funded study: "Ultimately, it must be the government's responsibility to 'make life easier' for game developers..." (SECOR 2008, 36).

New media, a major subset of creative industries, has a long-standing association with entrepreneurial self-reliance and "anti-corporate" work cultures. The post-2000 expansion of Canada's digital play business is, however, attributable less to a go-it-alone business sensibility than state subsidy to massive multinationals. Paradigmatic is Ubisoft, a France-based multinational with twenty-five studios in seventeen countries. Prompted by a Quebec lobbyist's proposal, a mix of provincial and federal tax breaks enticed the publisher to set up in Montreal in the late 1990s (Tremblay and Rousseau 2005). Today, Ubisoft's Montreal studio is among the world's largest. At an estimated cost to the province of $100 million in 2010 (Stechyson 2011), Quebec's multimedia tax credits are reported to have attracted some eighty firms (Brodie 2011). Quebec has been a model for other jurisdictions, including Ontario, where Ubisoft opened a studio after that province promised $260 million in incentives. When calculating where to expand, government assistance is simply "part of the equation," says one Ubisoft executive (cited in Ebner 2009, n.p.). In 2010 Quebec bought a 4.5 per cent stake in Ubisoft, tightening the knot of state actors and game capital in Canada (Chung 2010).

Game industry associations describe gaming as a leading sector of the national economy—but one that is nonetheless "at risk" (ESAC 2010, 2). That combination has made pleas for subsidy compelling. For instance, the BC Interactive Task Force told the BC government in 2009 that the province's historically robust game industry was in a "precarious position," as publisher power was gravitating to jurisdictions offering lower taxes (cited in Ebner 2009, n.p.). BC capitulated, introducing a 17.5 per cent credit. Industry incentives have been controversial, however, with some developers concerned that current programs tend to cater to the biggest corporate players (SECOR 2011, 22). Although micro-studios are eligible for modest project and marketing funding through agencies such as Telefilm, the single largest forms of government support for game production in Canada are tax breaks and other incentives from which foreign-owned firms stand to save most. Ultimately, the nature of state support for game capital in Canada is illustrative of the economization of cultural policy, a process whereby public contribution to cultural production is governed by market criteria of investment, return, and, especially, job creation.

Producers

"What makes us stay here is the talent," says the head of Ubisoft Montreal (cited in LaSalle 2010, n.p.). "Talent," a close cousin of the creative-class concept, is a pervasive term in the discourse on the creative economy, and commentary surrounding the game sector is no exception. This is understandable as the industry's profit springs from the intelligence, imagination, and interactions of those devoting their workdays to producing video games. This cognitive labour is, however, a costly commodity. With industry salaries in Canada in the orbit of $70,000–80,000, payroll is a studio's single greatest expense. It is not surprising that the tax credits for which industry has lobbied mostly apply to labour budgets. Such subsidy is typically viewed as a de facto job creation and retention program. And, indeed, the country's game workforce has swelled, growing by nearly a quarter between 2006 and 2009 alone (Nowak 2010). Outdone only by the US and Japan, Canada's digital play business directly employs some 16,000 people (SECOR 2011, 6).

Quantitative measures like aggregate employment and an innocuous-sounding phrase like talent, tend, however, to gloss over actual conditions of game labour—a criticism levelled against the creative economy model in general (Banks and Hesmondhalgh 2009). This section looks at the flipside of the industry's work-as-play reputation as well as the promise of "indie" game development.

A consideration of the political economy of game labour must begin with what one subsidy-skeptic claims is a prevailing assumption among policymakers that "video game jobs are somehow more valuable than other jobs" (Brown 2011, n.p.). Game employment is regarded as high scoring on pay, skill, and productivity. Industry associations jockeying for tax breaks point out that game workers' relatively high pay injects into government coffers a lucrative personal income tax stream (Kyllo 2011a). The technical proficiency, managerial coordination, and artistic ingenuity demanded in the game-development process jibes with the emphasis in creative industry talk on high-skill knowledge work as a key to a national competitive edge. With regard to productivity, the fruits of game labour comply with one of the accumulation strategies that define the creative industries—the generation of financial value via intellectual property rights.

Work in creative industries is generally perceived as glamorous. A game job has the added allure of being paid to play. Game labour is, by all accounts, challenging, rewarding, and exciting. Yet the industry doesn't entirely live up to its work-as-play image. One Canadian studio owner admits the business has a "tarnished reputation for treating people badly" (cited in Gooderham 2010). Especially controversial is overwork—an issue that was exposed less by game workers than by their disgruntled partners. In 2004 the anonymous "EA Spouse" wrote a widely circulated blog post detailing endemic excessive hours and the devastating impact these have on game workers' personal lives (Dyer-Witheford and de Peuter

2006). In 2010, "Rockstar Wives" repeated the gesture.

Game studios, like the emotional labours sustaining those working within them, are profoundly gendered. It is estimated that women make up about 10 per cent of the development workforce in Canada (Orford 2011; Wong 2010). This imbalance is partly the result of a feedback loop by which the industry's most enthusiastic recruits are drawn from its most devoted consumer base. The interplay of leisure and labour does not conclude, however, when an avid gamer lands an industry gig. A study of Montreal developers highlights the extent to which game workers' leisure choices are informed by their jobs and, in turn, inform their jobs, pointing to the blurring of work and non-work time (Charrieras and Roy-Valex 2008).

Gaming may be a booming business in Canada, but this does not mean its workforce is protected from precarity, or employment-related insecurities. Of the varied manifestations of precarity, bluntest is job loss. Expensive to develop, a console game that flops in the market can be tantamount to a pink slip. Risk is acute for rookie studios whose fate hangs on a single project. Publishing giants do not necessarily offer greater job security. Notwithstanding the perception that it is recession-proof, the game industry suffered deep job cuts in the wake of the economic downturn beginning in 2008. EA hemorrhaged 2,500 positions—10 per cent of its global workforce.

Vancouver studios that had been acquired by foreign-owned publishers were particularly vulnerable. For example, in 2010 Activision Blizzard laid off about 100 developers from Radical Entertainment, one of Vancouver's first studios. In 2011, Disney Interactive shut Propaganda Games, a seventy-person studio founded by former EA Canada developers. Significantly, some of the new studios sprouting up in Vancouver are responding to a risky market by substituting short-term, project-based contract workers for long-term studio employees (Kyllo 2011b), thereby making precarious employment a more formal part of the business model of game development. Some insiders anticipate the game sector moving to a freelance labour economy like that of film production (see Gouglas et al. 2010, 22–23). This model becomes more feasible as the "wealth of talent" grows (Williams cited in Kyllo 2011b, n.p.).

Game workers may also confront precarity in the form of fear of job loss due to outsourcing. Studios are exporting an increasing portion of development work to lower-wage locations such as China, India, and Vietnam. Outsourcing has been steadily inching up the value chain, and now affects everything from programming to animation. In the creative economy model, the source of a country's competitiveness is a creative class continually sharpening its expertise for those steps in a production process where the greatest return lies. Geographic diffusion of game-making know-how suggests that countries of the Global North are not guaranteed a monopoly in this regard. By the same token, it is important to recognize that Canada is itself a premiere outsourcing destination in the global game industry, with the country's biggest corporate players headquartered elsewhere.

A celebrity designer like Nintendo's Shigeru Miyamoto is the quint-essential creative-class worker. Games are, however, the product of a stratified workforce. In a study of the Vancouver game industry, the term "undercreative class" was proposed to designate the most precarious segments of game labour (Barnes et al. 2009). Highlighted here are those who play-test games for bugs: "testers" are the most likely to earn the lowest company wage, to be hired on an as-needed basis on a temporary contract, and to be excluded from perks available to full-time developers. Gaming's undercreative class encompasses those performing "free labour" as well (Terranova 2001). Gamer activity that contributes to the generation of financial value but is nonetheless unpaid has been theorized as "playbour" (Kücklich 2005). This hybrid activity spans from fan-made machinima, which can promote interest in a game, to pay-to-play MMOs, the attract-iveness of which is sustained by user-generated content.

Publishing giants and tax credits are not the lone forces of game indus-try growth in Canada. Developers' desire for greater creative experimenta-tion than that allowed at large studios has also driven expansion, particu-larly in Vancouver (Dyer-Witheford and Sharma 2005). No less than an EA executive has spoken openly about "creative failure" (cited in Androvich 2008, n.p.) in a sector that reaps enormous profit from a rinse-and-repeat cycle of annually updated sport games. Passion for a fresh game concept, entrepreneurial gusto, frustration with game projects being canned mid-way after a corporate buyout, and habituation to employment instabil-ity—these are among the factors in play when developers abandon their jobs at major studios to riskily launch a startup (Kuchera 2011).

These independent game developers sometimes challenge dominant design patterns. Vancouver's Silicon Sisters, for example, is a female-owned studio where games are made by women for women and girls. Lower production costs for mobile and social games, coupled with digital distribution platforms beyond publisher-controlled channels, hold out the promise of a renaissance in "indie" game development where the ownership of intellectual property is retained by its creators (Taylor 2010). The new independents should not be romanticized, however. Not only is independent game development a "precarious business" (Gouglas et al. 2010, 17), but also the business objective may well be "build to sell" (Barnes et al. 2009)—in some cases, as it happens, to the same moguls its founders initially fled.

Places

Industry incentives, skilled workers, and public infrastructure are among the qualities making Canada attractive to game companies. From HB Studios in Lunenburg, Nova Scotia, to Microsoft in Victoria, British Columbia, game production spans the country. The industry has, however, an uneven geographic presence. Nearly 90 per cent of the game firms in Canada are split among British Columbia (31 per cent), Quebec (21 per cent), and Ontario (35 per cent) (Gouglas et al. 2010, 7). Within these provinces, game jobs concentrate in a trio of cities—Vancouver, Montreal, and Toronto. This section offers a glimpse of these urban hubs, each with distinct but interacting trajectories. It notes that the production of video games contributes to the transformation of place and, in the process, exacerbates some of the fault lines of the creative city.

Vancouver is the oldest of Canada's game clusters. The region employs some 3,500 game workers at about 140 companies (Kyllo 2009b). The studio setting off the Vancouver-area industry was Distinctive Software, founded in 1982 and acquired by EA in 1991. EA Canada's 500,000-square-foot campus in suburban Vancouver now staffs about 1,500 developers churning out top-selling console titles. When Distinctive was bought, some of its developers parted ways to form Radical Entertainment. This became a familiar force of industry expansion in Vancouver: "a group of talented developers who have gained experience working in a larger company decide to split off and follow their own creative vision" (*ibid.*, n.p.). EA and Radical unwittingly seeded a chain of spinoffs: Barking Dog, Black Box, Propaganda, Relic, and many, many more. The network of studios that entrepreneurs have created is one of the few "safety net(s)" available to Vancouver developers who find themselves out of work in a volatile industry (Gouglas et al. 2010, 16).

Most Vancouver game firms are "within a few blocks" in and around Yaletown (Hickling Arthurs Low 2009, 7). This warehouse district was remade as Vancouver transitioned to a post-staples economy (Barnes and Coe 2011). Supplying the loft aesthetic commonly associated with new media work, Yaletown is "the most prestigious address for a range of specialised creative industries" (Barnes and Hutton 2009, 1260). Escalating rents priced out the very cultural-worker residents who contributed to the area's initial cachet. Catering to high-profiting tech sectors and the high-earning creative class they employ, Yaletown is dotted with condos, cafes, upmarket restaurants, and specialty shops. Gentrification has been paralleled by corporate consolidation, with several Vancouver upstarts swallowed by Disney, Take-Two, Vivendi, and other multinationals looking for a toehold in this globally recognized game development city. Vancouver's sector grew with relatively little in the way of industry-specific tax breaks. It was only in 2010 that BC introduced labour credits to compete with incentives available in jurisdictions such as Quebec.

Quebec has the largest provincial game labour force—three-quarters of which is in Montreal; one city booster says it is a potential "Hollywood of gaming" (cited in Leijon 2005, n.p.). Its nucleus is Ubisoft. Arriving in 1997, the France-based firm built its biggest studio—1,800 developers—in Montreal. A French-speaking populace and a workforce trained in digital animation made Montreal appealing. Most tempting, however, was an accommodating province that in the 1990s bet on multimedia as a path to post-industrial prosperity. Quebec wooed Ubisoft with a whopping 37 per cent tax break on labour costs. Ubisoft's hit *Splinter Cell*, the team that made it, and the tax deal that slashed its budget attracted an inflow of firms such as EA (US), Eidos (Japan), and Warner Bros. (US). While start-ups have focused on emerging platforms are proliferating, the majority of Montreal's developers work for large foreign-owned studios (Gouglas et al. 2010, 19). The city's vaunted talent pool is a hotly contested resource, with disputes over "talent poaching" between companies playing out in city court rooms (Leger 2011).

As in Vancouver, game companies have had "ground-level impact" on Montreal, particularly in Mile End (Nowak 2011b). At the "epicentre" of this area's transformation is the Ubisoft studio inhabiting a former industrial textiles building (Nowak 2010). Mile End, claims one of the studio's founders, was "just emptiness" before Ubisoft arrived in the late 1990s. "[N]ow," he reports, "it is so vivid, so lively" (cited in *ibid.*, n.p.). Such talk of urban regeneration neglects the lower-income residents displaced from a neighbourhood as it is remade by and for creative industries. And yet the "revitalizing effect" has been described as "precisely" the province's goal of luring foreign studios (*ibid.*). Indeed, the "Montreal model" of game-led urban economic development has been something of a template for other cities desiring a game cluster of their own.

Enter Ontario, which hosts Canada's third-largest game hub. A distinctive quality of the geography of game production in Ontario is its dispersion. Significant locally owned, mid-sized studios operate in small cities, such as Digital Extremes in London and Silicon Knights in St. Catharines. The Greater Toronto Area is, however, an incipient hub. Big brands like Capcom (Japan) and Rockstar (US) have Toronto satellites. Most notable, though, is the prevalence of small firms. While its game workforce is estimated to be a third the size of Montreal's, Toronto is home to a greater number of companies (Hickling Arthurs Low 2009, 13). Lacking access to a major employer like EA in Vancouver, Toronto developers have struck out on their own in droves, launching studios specializing in lower-cost platforms like hand-helds, mobiles, and console networks (e.g., DrinkBox Studios, Get Set Games, Queasy Games). The vibrancy of the city's indie game community is reflected in the annual Toronto Game Jam where developers create entire games on a single weekend. Toronto's indie scene has spawned star studios like Capybara Games, which arose from informal gatherings of the Toronto chapter of the International Game Developers Association.

The Ontario Liberal government wanted to get in on the game at another level. Learning from Vancouver and Montreal, the province began shopping for a major foreign-owned publisher that might serve as an "anchor tenant" (Ebner 2009). Following two years of courting, in 2009 the McGuinty Liberal government announced its deal with Quebec's darling Ubisoft. The next year it opened a Toronto studio to which the province offered incentives valued at $263 million on Ubisoft's commitment to create 800 jobs by 2020. The role of the Ubisoft Toronto studio in accelerating "gentrification" in its Junction neighbourhood is already being assessed (McGinnis 2010). One commentator expressed hope for the area "getting cleaned up" (Nowak 2011b), language that, once again, eclipses the possibility of community displacement and cultural homogenization following creative-class occupation.

Three points can be highlighted from this skeletal outline of Canada's game development clusters. First is the extent of provincial government participation in the inter-urban competition for game jobs. Provincial game-industry policies and investment partnerships are, moreover, increasingly promotional occasions unto themselves, place-marketing opportunities to communicate to international game companies that Canada is a destination to include in expansion plans. A second point concerns productivity. The internal revenue of Canadian game studios is not the only measure of economic contribution. Another is the property market to the extent that game companies stoke gentrification, and, in the process, heat up real-estate value. A third point concerns mobile capital. Canada is currently a choice location for game production—but game capital's mobility amplifies as the talent pool in other countries grows. Studio executives make no secret of "keep[ing] their eye on the business environment in other jurisdictions" (Chung 2010, n.p.). As an executive at the Montreal studio of Japanese-owned Eidos puts it: "Nothing is forever" (cited in *ibid.*).

Talent Nationalism

At the start of the twenty-first century, there is unprecedented enthusiasm invested in Canada's video game industry. Gaming, one journalist wagered, "could be Canada's best chance at establishing a strong cultural industry" (Seguin 2005, n.p.). Developers, too, have begun to narrate the sector as a "source of national pride…a creative industry that we can all be proud of" (cited in Kyllo 2011b, n.p.). Notably, however, the swell of industry self-studies, news items, and government publications boasting about Canadian game industry growth contain barely a reference to the content of the games themselves, let alone their national complexion. True, the cell of Vancouver's video game industry, *Evolution*, had a level featuring a beaver; *NHL* is indeed made in Canada; and a social game called *Pot Farm* did bud from a West Coast studio. In the main, though, the official discourse on video game production in Canada puts little stock in the content/identity couplet that was a trope of twentieth-century Canadian cultural policy. Where Canadian game development projects are frequently based on foreign-owned licensed properties, where foreign media ownership rules do not apply, and where Canadian-made games are targeted first to export markets, the "source of national pride" is located elsewhere.

It is a sharp turn of events: a medium born from a desire to escape work is today embraced by governments for its potential to create jobs, and thus satisfies a principal creative-economy imperative. While the federal government boasts of the "deep talent pool" awaiting foreign game companies in Canada (Invest in Canada 2011, 1), Canadian industry insiders have begun to warn of "talent risk"—a gambit, perhaps, of game companies eager to harness public-sector education institutions as their "talent … pipeline" (SECOR 2008, 20). From the intellectual property it generates to the urban districts it regenerates, the game workforce is, for the moment, a cherished economic resource in Canada, albeit one that displays a familiar geographic pattern of uneven development. If the building of earlier transport and broadcast systems in Canada were products and projects of "technological nationalism" (Charland 1986), then the discourse on the country's video game industry is suggestive of talent nationalism, or the identification of sovereign prowess with cognitive labour capacity in an age of globalizing, hyper-competitive digital capitalism.

References

Afan, E. 2010. "Canada Poised to be a World Leader in Digital Economy." *Playback* (May), 45.

Androvich, M. 2008. "D.I.C.E. 2008: Says Riccitiello." *Gamesindustry.biz*, August 2. Accessed November 3, 2011. http://www.gamesindustry.biz.

Banks, M., and Hesmondhalgh, D. 2009. "Looking for Work in Creative Industries Policy." *International Journal of Cultural Policy* 15 (4): 415–430.

Barnes, T., and T. Hutton. 2009. "Situating the New Economy: Contingencies of Regeneration and Dislocation in Vancouver's Inner City." *Urban Studies* 46 (5–6): 1247–1269.

Barnes, T., and N. M. Coe. 2011. "Vancouver as Media Cluster: The Cases of Video Games and Film/TV." In *Media Clusters: Spatial Agglomeration and Content Capabilities*, ed. C. Karlsson and R. Picard, 251–277 Northampton, MA: Edward Elgar.

Barnes, T., A. Holbrook, T. Hutton, and R. Smith. 2009. "Creativity and Innovation in the Vancouver City-Region." Paper prepared for Meeting of the Innovation Systems Research Network, Toronto, Ontario, November 5–6.

Beck, J., and M. Wade. 2004. *Got Game: How the Gamer Generation is Reshaping Business Forever.* Boston: Harvard Business School Publishing.

Brodie, T. 2011. "Quebec Luring Video Game Jobs from U.S." *Globe and Mail*, September 6. Accessed November 16, 2011. http://www.theglobeandmail.com.

Brown, J. 2011. "Grand Theft Tax Break." *Maclean's*, September 13. Accessed November 16, 2011. http://www2.macleans.ca.

"Canada World's Third Largest Video Game Producer." 2011. *Calgary Beacon*, March 2. Accessed October 23, 2011. http://www.calgarybeacon.com.

Castronova, E. 2005. *Synthetic Worlds: The Business and Culture of Online Games.* Chicago: University of Chicago Press.

Charland, M. 1986. "Technological Nationalism." *Canadian Journal of Political and Social Theory* 10 (1): 196–220.

Charrieras, D., and M. Roy-Valex. 2008. "Video Game Culture as Popular Culture? The Productive Leisure of Video Game Workers of Montreal." Paper presented at the Annual Meeting of the International Communication Association, Montreal, Quebec.

Chung, E. 2010. "Video Game Subsidy Battle Heats Up." *CBC News*, September 14. Accessed November 16, 2011. http://www.cbc.ca/news.

Cross, T. 2011. "All the World's a Game." Special Report: Video Games, *The Economist*, December 10.

Dyer-Witheford, N., and G. de Peuter. 2006. "'EA Spouse' and the Crisis of Video Game Labour: Enjoyment, Exclusion, Exploitation, Exodus." *Canadian Journal of Communication* 31(3): 599–617.

Dyer-Witheford, N., and Z. Sharma. 2005. "The Political Economy of Canada's Video and Computer Game Industry." *Canadian Journal of Communication* 30 (2): 187–210.

Ebner, D. 2009. "Toronto Scores Points in the Video Game Sector." *Globe and Mail*, August 3. Accessed December 18, 2011. http://www.theglobeandmail.com.

Entertainment Software Association of Canada (ESAC). 2010. *Game On, Canada!* Toronto: Entertainment Software Association of Canada. Accessed October 25, 2011. http://www.theesa.ca.

———. 2011. *2011 Essential Facts About the Canadian Computer and Video Game Industry.* Entertainment Software Association of Canada. Accessed October 25, 2011. http://www. theesa.ca.

Florida, R. 2003. *The Rise of the Creative Class: And How It's Transforming Work, Leisure, Community and Everyday Life.* New York City: Basic Books.

Gooderham, M. 2010. "Career Satisfaction—and a Life." *Globe and Mail*, January 10. Accessed June 9, 2012. http://www.theglobeandmail.com.

Gouglas, S., et al. 2010. *Computer Games and Canada's Digital Economy: The Role of Universities in Promoting Innovation.* Report to the Social Sciences and Humanities Research Council, Knowledge Synthesis Grants on Canada's Digital Economy.

Hickling Arthurs Low. 2009. *Canada's Entertainment Software Industry.* Ottawa: Hickling Arthurs Low Corporation.

Hon, D. 2009. "Rough Play in Vancouver Video Games." *BC Business*, June 3. Accessed November 16, 2011. http://www.bcbusinessonline.ca.

Invest in Canada. 2011. *Digital Media: Canada's Competitive Advantages*. Ottawa: Foreign Affairs and International Trade Canada. Accessed December 19, 2011. http://investincanada.gc.ca/download/839.pdf.

Kuchera, B. 2011. "Feral Developers: Why Game Industry Talent is Going Indie." *Ars Technica*. Accessed December 19, 2011. http://arstechnica.com.

Kücklich, J. 2005. "Precarious Playbour: Modders and the Digital Games Industry." *Fibreculture Journal* 5. Accessed October 18, 2011. http://journal.fibreculture.org.

Kyllo, B. 2009a. "Looking for a New Way to Invest? Make a Game of It." *Globe and Mail*, March 12. Accessed December 19, 2011. http://www.theglobeandmail.com.

———. 2009b. "Vancouver's Video Game Family Tree." *Georgia Straight*, January 29. Accessed November 24, 2011. http://www.straight.com.

———. 2011a. "Vancouver Video-Game Industry Still Suffering." *George Straight*, March 29. Accessed November 24, 2011. http://www.straight.com

———. 2011b. "Vancouver Video-Game Industry in Transition." *Georgia Straight*, September 1. Accessed November 24, 2011. http://www.straight.com.

Landry, C. 2008. *The Creative City: A Toolkit for Urban Innovators*. 2nd edition. London: Earthscan.

LaSalle, L. 2010. "Canada Is among Top Video Game Developers with Its Home-grown Titles." *Globe and Mail*, July 16. Accessed October 12, 2011. http://www.theglobeandmail.com.

Leger, K. 2011. "New Court Knock-Out in Montreal Video-Game Talent Poaching Fight." *The Gazette*, December 23. http://blogs.montrealgazette.com.

Leijon, E. 2005. "Game On." *Montreal Mirror*, October 27–November 2. Accessed November 23, 2011. http://www.montrealmirror.com.

McGinnis, R. 2010. "Does Ubisoft's Arrival Signal Gentrification for the Junction Triangle?" *blogTO*, October 14. Accessed December 19, 2011. http://www.blogto.com/.

McGonigal, J. 2011. *Reality Is Broken: Why Games Make Us Better and How They Can Change the World*. New York: Penguin.

Nowak, P. 2010. "Respawned: How Video Games Revitalize Cities." *CBC News* (September 14). Accessed December 19, 2011. http://www.cbc.ca/news.

———. 2011a. "Canadian-made Games and the Question of Outsourcing." *CBC News* (February 4). Accessed October 14, 2011. http://www.cbc.ca/news.

———. 2011b. "In Praise of Video Game Subsidies." *Maclean's*, September 15. Accessed October 14, 2011. http://www2.macleans.ca.

Orford, S. 2011. "Wanted: More Women in Vancouver's Video-Game Industry." *Georgia Straight*, September 7. Accessed November 24, 2011. http://www.straight.com.

SECOR. 2008. *Ontario 2012: Stimulating Growth in Ontario's Digital Game Industry*. Toronto: SECOR Consulting Ltd.

———. 2011. *Canada's Entertainment Software Industry in 2011*. Toronto: SECOR Consulting Ltd.

Seguin, D. 2005. "Search and Employ." *Canadian Business*, April 1l. Accessed October 16, 2011. http://www.canadianbusinessonline.com.

Stechyson, N. 2011. "Province Lagging in Video Game Industry." *Telegraph-Journal*, September 8. Accessed December 18, 2011. http://www.telegraphjournal.canadaeast.com.

Takahashi, D. 2011. "Modern Warfare 3 Sells $1B in Just 16 Days." *GamesBeat*, December 12. Accessed December 12, 2011. http://www.venturebeat.com.

Taylor, K. 2010. "Canada's Video Game Industry is a Going Concern." *Toronto Star*, September 24. Accessed Octover 16, 2011. http://www.thestar.com.

Terranova, T. 2001. "Free Labor: Producing Culture for the Digital Economy." *Social Text* 18, 2 (63): 33–58.

Tremblay, D-G., and S. Rousseau. 2005. "The Montreal Multimedia Sector: A Cluster, a New Mode of Governance or a Simple Co-location?" *Canadian Journal of Regional Science* 28 (2): 299–328.

Wong, W. 2010. "Women Missing from Video Game Development Work Force." *Chicago Tribune*, August 5. Accessed November 24, 2011. http://chicagotribune.com.

5

Book Publishing: Dying One Chapter(s) at a Time?

Jeff Boggs

This chapter surveys changes in the Canadian book trade since 1996. While using the second edition of *The Cultural Industries in Canada* as a baseline, this chapter focuses on changes in book distribution and retailing as the most important forces shaping the industry. Three commonalities remain. First, the Canadian book trade still consists of an anglophone and a francophone market. As with the 1996 edition, this chapter focuses on the English-language book trade. Second, technological change continues, shifting the competitive terrain and complicating attempts to nurture a Canadian-controlled book trade. The resulting division of labour adds some new tasks and organizations, eliminates some, and transforms others. Third, cultural nationalist policies still influence the competitive environment. Legislators assumed that the sole source of problems would be multinational foreign competition, but the position of foreign-controlled publishers has slightly worsened. However, policy-makers did not anticipate the rise of the mega-bookstore Chapters or the online book trade of the Internet and how these would affect the industry. While changes in English-language Canadian book publishing were modest in production and distribution, changes in retailing generated a book trade barely imaginable fifteen years ago.

Content Production

Content production is concentrated in two provinces: Ontario and Quebec. In 2009, Ontario accounted for 62 per cent of operating revenue, and Quebec, 32 per cent; publishers in these provinces accounted for 98 per cent of operating profits (Minister of Industry 2011). Toronto dominates English-language book publishing whereas Montreal serves a similar role for francophone book publishing (Boggs 2010). Nationally, there are approximately 1,500 book publishers, with an estimated $2 billion in revenue and 9,000 workers concentrated in the top 300 or so businesses (Department of Canadian Heritage 2010).

While the major patterns in content production are long established, specifics are painstaking to ascertain, largely due to changes in how Statistics Canada collects and releases data. For instance, Canada Catalogue

Table 5.1 Book Publishing Sales, 2008

Sales in Canada of all titles	$1,528,412
...by Canadian-controlled establishments	$803,626 (53%)
...by foreign-controlled establishments	$724,786 (47%)
Domestic sales of own-titles	$939,880
...by Canadian-controlled establishments	$617,975 (66%)
...by foreign-controlled establishments	$321,905 (34%)
Agency sales	$588,533
...by Canadian-controlled establishments	$185,652 (32%)
...by foreign-controlled establishments	$402,881 (68%)
Export and other foreign sales	$233,292
...by Canadian-controlled establishments	$220,325 (94%)
...by foreign-controlled establishments	$12,957 (6%)
Operating profit	$201,522
...for Canadian-controlled establishments	$84,997 (42%)
...for foreign-controlled establishments	$116,525 (58%)
Corresponding profit margin	9.7%
...for Canadian-controlled establishments	7.2%
...for foreign-controlled establishments	13.1%
Salaries, wages, and benefits	$397,503
...for Canadian-controlled establishments	$238,902 (60%)
...for foreign-controlled establishments	$158,601 (40%)

NB: All dollar values in current thousands. Sales data collected from 195 establishments accounting for 98 per cent of sales in 2008. Percentages correspond to share of major category and are rounded to two decimals.
Source: Minister of Industry (2010).

87-210, upon which Rowland Lorimer (1996) drew, was discontinued in 1995. The replacement series began about 1996 and was modified in 2002 (both in Catalogue 87-0004XIE), further complicating comparisons with Lorimer's work, as some indicators—most importantly, numbers of workers and titles—were dropped. The current data, however, still have some utility in identifying broad trends.

Table 5.1 provides an overview of sales data for book publishers in Canada. The 195 establishments surveyed in 2008 generated $1.528 billion in sales. Slightly more than half the sales—53 per cent—were captured by Canadian-controlled establishments, which has been fairly consistent for at least the last decade. These data obscure, however, that foreign-controlled sales are made by less than twenty establishments (cf. Boggs 2010). Canadian-controlled establishments generate two-thirds of all own-title sales but less than one-third of agency sales and nearly all export sales. This underlines the revenue streams for Canadian-controlled vs. foreign-controlled publishers: own-title sales; agency sales; and rights. Agency sales and agency publishing arise from being the sole Canadian distributor of a title owned by foreign publishers. Own-title sales involve the sale of a house's own titles, with higher costs and risks. Rights refer to the sale of publishing rights to another copyright jurisdiction such as the US or France. Thus, foreign-controlled establishments, while paying 40 per cent of wages and capturing just under half the sales, garner 58 per cent of the total profit. This is also seen in the higher profit rate for foreign-controlled publishers.

A common assumption is that Canadian-controlled companies acquire and publish titles with smaller sales. Foreign-controlled publishers concentrate on

Table 5.2 Book Publishing Sales by Official Language, 2008

	English	French
Sales in Canada	$1,180,388	$348,024
...of own titles	$684,319	$255,560
...as exclusive agents	$496,069	$92,464
Exports and other foreign sales	$210,742	$22,550
Operating profit	$155,914	$45,609
Operating profit margin	9.6%	10.2%
Salaries, wages, and benefits	$324,603	$72,890

NB: All dollar values in current thousands.

Source: Minister of Industry (2010).

fewer titles, selecting those already popular in other markets. Furthermore, foreign-controlled publishers are better positioned—due to parent companies—to act as agency publishers. Until Statistics Canada or BookNet Canada releases their confidential data, these assumptions cannot be probed further.

Another set of differences in the Canadian context is the existence of two official languages in the same copyright regime. The numbers of speakers of each, not Canada's total population, set the maximum market size. Table 5.2 provides a breakdown of sales in these markets in 2008. All other things being equal, books in smaller markets tend to be more expensive than books in larger markets, as fixed costs are spread out over a smaller print run. Given that the Canadian market really consists of two book markets, these fixed costs are spread over an even smaller number of copies, driving up the cover price. However, despite the smaller market, operating profit margins are higher in the francophone market.

The agency system clearly relies on book imports. The data in table 5.1 acknowledges exports, predominately from Canadian-controlled publishers. Figure 5.1 reveals trends in exports and imports from 1996 to 2010, with one caveat. These data only capture goods that have crossed the Canadian border; downloaded or electronically transmitted goods are not included. Thus, more recent years are less reliable in capturing total exports and imports than earlier years. Canada's total book imports range

Figure 5.1 Total Book Imports and Exports, Plus Contributions of Largest English- and French-language Partners, 1996–2010.

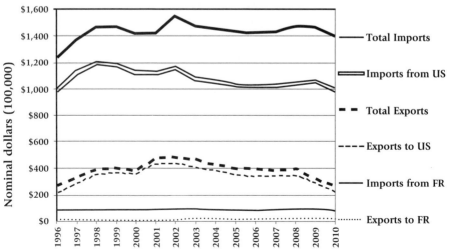

Source: Author's analysis of Minister of Industry (2006, 2011c).

between $1.2 and $1.6 billion, fluctuating with the exchange rate. The table also includes data for the US and France. Imports largely originate from the US, although a larger share of Canada's book exports are to the US. While the quantities are smaller for France, the same relationship exists: Canada imports more books than it exports.

Overall, these findings are similar to Lorimer's (1996), suggesting a holding pattern. The changes over fifteen years have been similar to changes since the 1970s. First, there have been changes in ownership. For instance, in 2002, Scott Griffin acquired Anansi Press (MacSkimming 2007). In 2000, Random House Canada acquired one-quarter of McClelland & Stewart, while the remainder of the publishing house was donated to the University of Toronto. This arrangement made M&S and its imprints an independent subsidiary of Random House Canada (Cyderman 2000), though some cultural nationalists saw this as a violation of rarely enforced foreign ownership regulations. In 2012, the University of Toronto sold all its ownership to Random House Canada (Quill 2012). Second, small houses continued to sign new authors who upon becoming successful tended leave for larger houses (e.g., Medley 2010). Third, the loss of an agency contract still sends Canadian-controlled publishers into a tailspin. For example, in 2009, Hachette ceased its agency relation with Key Porter, eliminating 35 per cent of Key Porter's revenue (Quill 2012). This meant that Key Porter, which published one of the largest Canadian non-fiction lists, had to close in early 2011. Fourth, sometimes Canadian-controlled publishing houses acquire rights to an international bestseller. Raincoast Publishing is the exemplar here: from 1999 to 2010, it held the Canadian rights to the *Harry Potter* franchise. Those profits underwrote the publication of Canadian-authored titles, and the loss of the *Harry Potter* titles led to the closing of Raincoast's publishing program and to a new focus on distribution.

These garden-variety changes, however, were accompanied by less-anticipated technological changes. At the root of these was the Internet. The emergence of online-only publishers has rendered Canadian rights less viable. Digital books can be bought from websites in other jurisdictions, reducing demand for titles in Canada's copyright space. Vanity publishers face competition from turn-key operations that enable self-publishing, allowing alternative channels for genre literature. Coupled with print-on-demand, this means that online-only publishers and turn-key operations can compete for readers who historically went through normal distribution channels, Canadian- or foreign-controlled (Davy 2007). They also face market incursions from Amazon.ca Books as it negotiates deals directly with authors. The size of its advances and ability to monitor self-published bestsellers on its own site make it a powerful new entrant.

Although enabled by technology, these changes do not operate in a policy vacuum. One could argue that the federal government is failing to enforce three sets of regulations shaping competition. First, the federal government fails to enforce rules on foreign ownership laid out in the *Investment Canada Act* (Williams 2011). Second, the Competition Bureau has allowed retail market concentration, enabling quasi-monopolies to demand concessions from publishers without passing the savings on to readers (Turner-Riggs 2007). "Pay-to-play" co-promotion policies make it difficult for smaller publishers to access shelf space (Castledale Inc. with Nordicity Group 2008). Third, the *Copyright Act* is ignored as retailers and distributors import books even when the territorial rights belong to a Canadian-based publisher. Be it *Harry Potter* (Eichler 2000) or more recent titles (Turner-Riggs 2007), each copy bought through parallel importation is one less purchased from the Canadian rights-holder, reducing the utility of Canadian copyright.

Moreover, irregular enforcement suggests waning support for cultural policies with roots in the 1970s. Eggleton (1997) argued that Canada's cultural policies needed revision, not just to comply with liberalized trading and investment rules, but also in response to new technologies. The 2010 Department of Canadian Heritage review of foreign investment adopts this position. In this light, the 2010 renaming of the DCH's Book Publishing Industry Development Program into the Canada Book Fund formalizes a transition already long underway, shifting resources from incubating domestic publishers to awarding subsidies to those with high sales, regardless of their questionable subcontracting practices and generous interpretation of "Canadian content" (e.g., Boggs 2010).

Wholesaling and Distribution

Lorimer (1996) paid scant attention to wholesaling and distribution, as the number of these operations is small. A recent study lists fifty in 2010, with most business being conducted by a handful (Department of Canadian Heritage 2010). These are warehousing operations, involving managing inventory, stocking orders, delivering titles, and managing accounts. Margins are thin and this segment has witnessed extensive restructuring since the 1990s.

As with publishing and retail, the Internet and enhanced logistic systems drive changes among distributors. These changes are intended to reduce costs and delays. Online retailing has impacted distribution in that

Canada's two largest online book retailers, Amazon.ca and Chapters.ca, have their own internal distribution centres. Each book ordered online equals one fewer flowing through the brick-and-mortar supply chain. Some multinationals now source their own titles from their US parents (Turner-Riggs 2008). Each book arriving in Canada through parallel importation means one fewer passing through the traditional supply chain. Often Ontario-centric, traditional distribution systems make it difficult to fill orders quickly in Canada's far-flung corners, and hence some retailers do engage in parallel importation from closer sources in the US (Taylor 2011). E-books and print-on-demand will also reduce the demand for traditional distributors. What book distribution looks like in the next edition of this book will depend on the federal government's enforcement of parallel importation regulations. Parallel importation, more so than foreign ownership, will ultimately result in most Canadian-owned distributors being replaced by more efficient US-based intermediaries. While this would be bad for the current Canadian-controlled distributors, it might be the best outcome for authors, publishers, readers, and even retailers.

Retailing: The Elephant in the Room

Of the three segments of the book trade, retailing has experienced the largest transformations and novel interventions. Where retail was once an assortment of brick-and-mortar bookstores, including a few small chains, it is now made up of one chain, a handful of independent bookstores, the new retailers Walmart and Costco, and the online presence of Amazon.ca Books. While books are still sold, their form and distribution channels are undergoing a transformation. Three interrelated trends stand out.

The first involves retail concentration. The number of independent booksellers declined alongside an increase in chains selling books, either as their primary offering via Chapters, or as a secondary offering via Walmart or Costco. However, brick-and-mortar chains are not the only source of competition.

The newest source of competition represents the second trend: online commerce. This takes two forms: online sales of bound titles and e-books. Collectively, Amazon.com and its Canadian subsidiary, Amazon.ca, captured 53 per cent (i.e., CAD$1.432 billion) of Canada's Internet retailing in 2010 (Euromonitor 2011a). Indigo's online operation netted 3.5 per cent of Canada's Internet retailing in 2010 (Euromonitor 2011b), though it accounted for 31 per cent of total retail value of media sales. Whereas Amazon.com—with 37.1 per cent of Canada's Internet retailing—sells consumer electronics, music, toys, and software in addition to books, Amazon.ca—with 15.7

per cent of Canada's Internet retailing—focuses on books, DVDs, and music (*ibid.*). These figures do not distinguish between bound books, e-books, and other goods. Still, they suggest that online operations are quite competitive.

Online commerce also takes the form of digital delivery of e-books. Lorimer (1996, 24–26) noted that CD-ROMs and the Internet provided opportunities for transforming reference books. However, the emergence of e-book readers (e.g., Kindle, Kobo, or Nook) and handheld computers (e.g., iPad, iPhone) signals new technology that might reduce demand for hard-copy books. As Amazon.ca is now selling more e-book titles than hard-copy, and with e-book sales having topped $1 billion at the end of 2010 (Barber 2010; Krashinsky 2011), perhaps this day is at hand. E-books might be "the killer app" of independent bookstores.

Independent bookstores rely on a range of industry-specific collective infrastructure such as trade associations, book prizes, and trade fairs, all aiding the circulation of information. In English-language publishing, the Canadian Booksellers Association (CBA) has provided such infrastructure since the early 1950s. Likewise, book prizes, including the Giller Prize, emerged in the 1990s to complement the older Governor General's Literary Award. Despite the survival of the CBA or the addition of new literary prizes, critical collective infrastructure supporting independent bookstores collapsed. This is the third trend. As chains and e-commerce capture more activity, independent bookstores and their contributions as members of trade associations, book fairs, and readers of industry periodicals wane. While Chapters' management hosted its 1999 annual sales meeting in the Bahamas to draw publishers away from the CBA fair (Ross 1999b), problems continued even after Chapters was taken over by the comparatively angelic Heather Reisman. The most telling example is the cancellation of the 2009 BEC (Book Expo Canada). Reed Expositions, Inc. closed it after Canada's largest houses announced they would no longer attend. Presciently, the BEC's previous organizers—the CBA—sold it to Reed Expo in 2000. After the BEC closed, the CBA launched a smaller fair focused as much on workshops and training for small retailers as on collecting orders.

The Catalytic Role of Chapters in Driving These Trends

When the second edition of this book was written, concerns about the possibility of a US-based chain wreaking havoc on the Canadian market were widespread. In fact, it was the domestic threat of the first iteration of Chapters that transformed the anglophone book trade. Chapters should be seen less as an innovator than as the first firm to be capitalized sufficiently to restructure Canada's book market. While cultural policy prevented foreign-controlled retailers from opening storefronts, it did not prevent Canadians from adopting the innovations of foreign retailers.

By the 1980s, book retailing in the US was being captured by ever-larger chains. Chains exploit scale economies by standardizing ordering, distribution, and management, resulting in higher efficiency and profitability. Increased profitability was initially achieved through better-stocked and better-organized stores driving less efficient bookstores into bankruptcy. Later, this process was repeated with smaller chains being bankrupted or purchased by larger chains.

Fears about the Department of Canadian Heritage's weak will toward preventing foreign brick-and-mortar retailers from unleashing these innovations on the Canadian book market were ultimately unfounded (Corcoran 1996; Trueheart 1995; Walker 1995), but these regulators overlooked domestic retailers' ability to do the same. In fact, they allowed Chapters to form out of the fusion of Coles and Smith Books, producing a single book retailer of 430 stores (Ross 1995). This monopoly so frightened independent bookstores that the Canadian Booksellers' Association lobbied Ottawa—in vain—against the merger. Afterwards, Chapters resigned from the CBA (*ibid.*). While this locked Chapters out of the CBA book fair, it did not block its access to books. The CEO of Chapters, Larry Stevenson, claimed that Chapters' real competitors were not independent bookstores but Costco, Walmart, and other discount retailers (Southworth 1999).

Like other discount retailers, Chapters located superstores in high-accessibility locations, often near competitors. This common retail strategy drives up rents for surrounding stores and creates a retail agglomeration allowing comparison shopping. However, Chapters added strategies specific to the book trade. Their stores incorporated comfortable browsing areas and coffee shops, making it easier to linger. Chapters purchased books on such a large scale that it could demand 45 per cent discounts on titles, instead of the standard 40 per cent (Walker 1995; Martin 2000). In 1998, the three-year moratorium Chapters signed with the Competition Bureau expired, allowing it to push for even steeper discounts (Cole 1999). Furthermore, Chapters' tardy payments and high book returns were treated with impunity (Livingston 2000). Many publishers and distributors acted as if Chapters was too big to upset (Martin 2002), fearing that Chapters would not carry their titles (McNish and Strauss 2001).

However, Chapters botched implementing innovations in distribution, wholesaling, and online retailing (Martin 2002). In May 1999, a new wholesaler—Pegasus—was launched as a majority-owned holding of

Chapters (Canadian Press 1999). Publishers and independent booksellers alike eyed this development warily (Ross 1999a). Now Canada's largest retailer owned the largest book distributor, a level of vertical integration never seen in Canada's book trade. After a letter-writing campaign by publishers and CBA members (Ross 1999b) and following the US Federal Trade Commision's blocking of Barnes and Noble's acquisition of US book wholesaler Ingram (Lorinc 1999a), the Competition Bureau opened an investigation (Stoffman 1999).

By the time of the hearings, Pegasus had been demanding the 50 per cent discount given to wholesalers in the run-up to the Christmas retail season (Lorinc 1999b). It forced compliance by withholding orders until publishers agreed to the new terms (Lorinc 1999c). Despite these allegations, the Competition Bureau ultimately found no wrongdoing. It is unclear if this finding was influenced by publishers' fears of retaliation (Scoffield 2000).

Another misstep involved Chapters.ca, an online subsidiary of Chapters. Chapters did not anticipate online retailing when Smith Books and Coles were merged, and like many dot-com companies, Chapters.ca faced high start-up costs. While spun off as a publicly traded subsidiary to limit losses and capture new funding (Strauss 1999), this was badly timed with the dot-com crash. One would still expect "synergies" to emerge between the combination of these elements but this was complicated by competition from Amazon.com.

One would expect Chapters.ca and Chapters to be the largest customers of Pegasus. While this was the case, numerous anecdotes suggest that such synergies were difficult to realize (Strauss 2000). Despite centralized inventory and ordering, Chapters stores were still ordering books directly from publishers (MacSkimming 2007).

Collectively, these mistakes lost money and alienated the rest of the book trade, the Competition Bureau, and the Department of Canadian Heritage (Strauss 1999). Unfortunately for independent booksellers, consumers continued to flock to Chapters, Amazon.com, and Indigo. Despite an increase in its market share, bad press, rapid expansion, distribution problems, and the dot-com bust contributed to Chapters' falling to $9 per share in November 2000 (Strauss 2000) from $35 in mid-1999 (Pigg 2000). This loss paved the way for Indigo's $121 million takeover.

The change of ownership still found book publishers reeling from Chapters' late payments and high returns. In this environment, Indigo's CEO Heather Reisman laid out a code of conduct to allay publishers' fears. Independent bookstores—whose numbers fell by 15 per cent between

1995 and 2001—were happier facing just one book chain (Scoffield 2001). However, having dealt with a brick-and-mortar giant before, the industry demanded the Competition Bureau to weigh in. After the merger's first year, Chapters was required to pay for books within ninety days of possession and cap returns at 30 per cent (Pitts 2002). Indigo continued some practices instituted by the old Chapters. One example was "co-op promotion," which "allowed" publishers to pay for prime space on a table or on a bookshelf at the end of an aisle (or endcap) (Turner-Riggs 2007). Given that books so displayed have considerably higher sales, it was in the publishers' best interest to co-operate, if they could afford it.

Since the merger, Indigo's 240+ retail stores have continued positioning themselves as "cultural department stores" and are now restructuring floor space to capture non-book markets. For example, many stores reduced square footage in adult books to expand children's books, toys, and games in time for the 2011 holidays. This suggests that Indigo is losing the heightened competition for readers and is focusing more on "lifestyle products" (Euromonitor 2011c). Interestingly, Indigo's 2009 engagement with e-bookselling with the launch of its Kobo e-reader (*ibid.*) was short-lived; Indigo sold the reader in late 2011 for USD$315 million to Japanese online retailer Rakuten, Inc. (Strauss 2011).

Since entering the Canadian market in 2002 (Strauss 2002), Amazon.ca has added a new distribution facility and new product lines (Euromonitor 2011b). Along with independent booksellers, it provides additional competition. While the entrance of e-books by Google and others may create new book buyers, these technologies will probably capture sales that previously would have been from brick-and-mortar shops. Independent stores continue to survive, especially when they are able to find a niche market (Castledale Inc. with Nordicity Group 2008; Turner-Riggs 2007).

Collective Infrastructure

Canada's book trade is rife with collective infrastructure, such as trade associations, Copyright Canada, the Canada Council, the Writers' Union, the Canadian Book Fund, and so on. Collective infrastructure enables the industry to share information, identify best practices, and confront problems. One problem revealed in the late 1990s was the need for better inventory tracking. Prior to Chapters' emergence, it was accepted that filling an order might take weeks or months. Likewise, the inventory held by smaller chains was less than that held by Chapters, and with fewer returns. Returns at Chapters, despite an allegedly centralized warehousing and distribution centre, ranged over 50 per cent for some titles, a far cry from the

20 to 25 per cent of the early 1990s (Lorinc 2002). This created a range of problems for publishers.

The central problem was the lack of sales data until months after the sale. Fundamentally, this is because of the industry's consignment-like sales system. Publishers sell titles directly to retailers, or through a distributor or wholesaler, via a catalogue. Retailers order books from the catalogue for their shelves. Customers purchase some of these books. Unsold titles are returned, and these are deducted from what the retailer owes. Customarily, retailers have up to a year to return a title.

An example illustrates the difficulties. A publisher delivers 5,000 copies of a title to a retailer. Months later, the publisher assumes all copies have been sold, and pays royalties to the author. However, unbeknownst to the publisher, few of the books sold. One day close to the end of the return time, the retailer returns 4,000 unsold copies. Suddenly the publisher is faced with a conundrum, having paid royalties for 4,000 copies that never sold. The problem is further compounded if the publisher ordered a second printing in anticipation of more sales.

Before 1995, these outdated inventory control practices were mitigated by the fact that small independent bookstores never ordered—or lost—15,000 copies of a title, and then ordered 10,000 more copies, a problem experienced by McClelland & Stewart (McNish and Strauss 2001). Allegedly, in 2000, Chapters ordered 10,000 copies of Margaret Atwood's *The Blind Assassin* from McClelland & Stewart. According to the publisher's records, Chapters already had ordered 15,000 copies. After three days of phone calls, the original 15,000 were found in the Pegasus warehouse.

Before the Indigo and Chapters merger and the related bankruptcy of General Publishing, these problems never emerged on such a scale. Publishers tolerated the system because it allowed some latitude in which titles to publish. Retailers benefited in that they could return unsold titles, externalizing the risk to the publisher. After the first iteration of Chapters, Canada's publishers wanted a better system for tracking inventory (Avri 2003). Given the size of the retail market and its inability to house more than one major chain, it made more sense to increase efficiencies by updating the industry's system of tracking and warehousing books (Stoffman and Ziegler 2001).

In late 2002, BookNet (hereafter, BNC), an initiative between the Department of Canadian Heritage and the book trade, emerged as a collective solution to the problem (Himmelsback 2004). A non-profit funded from a fee (0.15 per cent) assessed on publishers' sales (Caldwell 2005), it uses point-of-sale (POS) data to track sales at each location (Avri 2003). It went online in 2005 (Atkinson 2005) and has continued to add not only subscribers—book publishers, independent book retailers, Indigo, big-box retailers—but a range of features. These features, however, rely on the subscriber adopting inventory-tracking software and bar-code scanning procedures (Caldwell 2005). For retailers making the investment, publishers are able to track real-time sales by customers, while independent retailers are able to monitor sales. Multiple store operations such as Indigo or Costco are better able to track sales geographically.

BNC ensures the timely flow of information between publishers, distributors, and retailers. It compiles and releases sales information so that publishers and retailers know how each title is selling. This transparency is crucial to the flow of information in a market too small for a private intermediary. BNC may more efficiently serve the industry than would a private monopoly. BNC now estimates it represents 75 per cent of sales in trade books, excluding textbooks, from over 1,500 retailers (BookNet 2010).

BNC appears to have reduced returns, given the absence of complaints in the trade and popular press. And unlike a national monopolist, it works in the best interest of its subscribers to upgrade its infrastructure and technical standards (e.g., Smith 2007; Canada News Wire 2007). It has also been able to deal with fears about proprietary or competitive information. This also explains why third parties, such as scholars, have a difficult time acquiring BNC data. Because BNC does not make raw data publicly available, it is hard to confirm that BNC is doing an effective job. This is another example of what Wagman (2010) sees as an opaque and many-armed cultural industry.

BNC does publish internal reports, however sanitized. Written as a means of helping the industry understand its sales, these reports also provide more reliable data than were previously available. For instance, the reports provided top-twenty lists of perennial bestsellers from 2006 to 2009 in fourteen categories to help retailers identify the best titles to stock (BookNet 2010). This shows that literary awards are associated with increased sales, at least for the 2006 and 2007 winners of literary prizes (BookNet 2009). Unfortunately, that report provides only a line graph for each award. While it is easy to see that winners of prizes have higher sales, no data are given on number of units sold. Thus, it is unclear if winning the Governor General's Award leads to higher sales than some other prizes. It also makes it difficult to determine if winning a foreign prize leads to higher sales than a domestic prize. And often, buried in the data, lurks the fact that most bestsellers are not written by Canadians for Canadians (Gesell 2006) but by foreign authors. Some might suggest this is troublesome, given that the industry receives millions of dollars a year in governmental support.

Conclusion

The central changes in Canada's book trade between 1996 and 2011 resulted from a combination of retail concentration and Internet-enabled business practices. Despite decades of subsidies, Canada's book trade remains shaky—as it does in most countries. Canada differs in that the book trade mobilizes cultural nationalism in its defence. However, this does not block the diffusion of new business practices; it merely means that only Canadian-controlled firms can implement them and thereby transform the industry.

References

Atkinson, N. 2005. BookNet Canada. *Publishers' Weekly*, June 13, 2005, S12.

Avri, J. 2003. "Booking on a Healthy Supply Chain: The Canadian Book Industry Is Faced with Great Challenges to Improve How It Does Business. Where Do Printers Fit into Its Future Plans?" *PrintAction* 33 (3): 24–26.

Barber, J. 2010. "Ebook Sales Are Close to $1-billion. From Sentimentalists to Imperfectionists, John Barber Explores How the Publishing World Shifted In 2010. The Same Old Story? The Novel Is Dead...and Yet." *Globe and Mail*, December 30, R1.

Boggs, J. 2010. "An Overview of Canada's Contemporary Book Trade in Light of (Nearly) Four Decades of Policy Interventions." *Publishing Research Quarterly* 26 (1): 24–45.

BookNet. 2010. "Perennial Bestsellers: Which Books Are Always at the Front of the Pack? A Category-Based Study of Titles with Consistently High Rankings Every Year from 2006 to 2009." In *BNC Research*: BookNet Canada. Accessed November 28, 2011. http://booknetcanada.ca.

———. 2009. "Literary Awards in Canada: What If You're Nominated? What If You Win?" In *BookNet Research*: BookNet Canada. Accessed November 28, 2011. http://booknetcanada.ca .

Caldwell, R. 2005. "Publishers Want to Know What We're Reading." *Globe and Mail*, May 2, R4.

Canada News Wire. 2007. "BookNet Canada Launches GDSN Pilot for North American Publishers." *Canada NewsWire*, June 5, 1.

Canadian Press. 1999. "Chapters Launches Book Wholesaler." *Globe and Mail*, May 7, D6.

Castledale Inc., in association with the Nordicity Group Ltd. 2008. *A Strategic Study for the Book Publishing Industry in Ontario*, ed. OMDC. Toronto: Ontario Media Development Corporation Book Industry Advisory Committee.

Cole, T. 1999. "Larry's Party: Larry Stevenson's Chapters Colossus Is the Future of Bookselling. Thing Is, No One Else Is Invited." *Globe and Mail*, September 24, 44.

Corcoran, T. 1996. "Living in a Borderless World." *Globe and Mail*, February 17, B2.

Cyderman, C. 2000. "Sale of Book Firm's Shares Raises Foreign-Control Fears." *Ottawa Citizen*, June 28, E1.

Davy, D. 2007. "The Impact of Digitization on the Book Industry." Toronto: Association of Canadian Publishers.

Department of Canadian Heritage. 2010. *Investing in the Future of Canadian Books: Review of the Revised Foreign Investment Policy in Book Publishing and Distribution*. Ottawa: Department of Canadian Heritage.

Eggleton, A. 1997. "Our Culture, Too, Must Compete in Global Marketplace." *Canadian Speeches* 10 (10): 3–6.

Eichler, L. 2000. "Raincost in Row with Toronto Indie." *Publisher's Weekly*, August 28, 18.

Euromonitor. 2011a. "Amazon.com - Retailing - Canada." In *Euromonitor International: Local Company Profile*, 2: Euromonitor Global Market Information Database.

———. 2011b. "Internet Retailing – Canada." In *Euromonitor International: Country Sector Briefing*: Euromonitor Global Market Information Database.

———. 2011c. "Indigo Books & Music Inc. – Retailing – Canada." In *Euromonitor International: Local Company Profile*: Euromonitor Global Market Information Database.

Gesell, P. 2006. "New Book-sales Tracker Reveals Canadians Not So Patriotic." *Postmedia News*, January 13, 1.

Himmelsback, V. 2004. "Canadian Book Industry Sorts Through Supply Chain." *Computing Canada* 30 (17): 25.

Krashinsky, S. 2011. "Amazon E-book Sales Surpass Paper." *Globe and Mail*, May 20, B9.

Livingston, G. 2000. "Questions Grow on Future of Book Business." *Toronto Star*, August 14, D1.

Lorimer, R. 1996. "Book Publishing." In *The Cultural Industries in Canada*, ed. M. Dorland, 3–34. Toronto: James Lorimer & Company Ltd., Publishers.

Lorinc, J. 1999a. "Dispute with Publishers Sparks Probe: Competition Bureau Looking into Chapters' Wholesale Subsidiary, Pegasus." *Globe and Mail*, September 21, C1.

———. 1999b. "Book Price Hike Seen by Next Fall." *Globe and Mail*, October 29, C16.

———. 1999c. "Federal Regulators Plan to Keep an Eye on Chapters Bookstore Chain: Competition Bureau Officials Say No Full-scale Inquiry Needed, but That They Have 'Concerns' about Some of the Bookseller's Practices." *Globe and Mail*, November 18, C5.

———. 2002. "Heather's Version: She Dominates Canada's Book Business – But Doesn't Need the Work." *Globe and Mail*, September 27, 42.

MacSkimming, R. 2007. *The Perilous Trade: Book Publishing in Canada, 1946–2006*. Toronto: McClelland and Stewart.

Martin, S. 2000. "Do We Need More Competition or Less? Should We Be Protecting the Sellers of Art, as Opposed to Its Makers?" *Globe and Mail*, December 4, R7.

———. 2002. "The Delusional World of Canadian Book-selling." *Globe and Mail*, May 15, R1.

McNish, J., and M. Strauss. 2001. "A New Chapter Opens in Publishing, PART 2: Code to Reduce Dependence on Giant Book-selling Chain." *Globe and Mail*, May 15, B1.

Medley, M. 2010. "Turnover of Authors a Fact of Life for Farm Teams of Can Lit." *Edmonton Journal*, March 26, D13.

Minister of Industry. 2006. "Culture Goods Trade: Data Tables 1996-2005." Catalogue #87-007-XIE. Ottawa: Statistics Canada.

———. 2010. "Book Publishers 2008." Catalogue # 87F0004X. Ottawa: Statistics Canada.

———. 2011. "Culture Goods Trade: Data Tables 2010." Catalogue # 87-007-x2011001. Ottawa: Statistics Canada.

Pigg, S. 2000. "Chapters Tries to Thwart Bid with Buyback." *Toronto Star*, December 9, D4.

Pitts, G. 2002. "Indigo CEO Girds for Series of Struggles; Book Retailer Faces Closings, Cost Cutting." *Globe and Mail*, January 7, B1.

Quill, G. 2012. "Canadian Book Publisher Key Porter Shuts Down." *Toronto Star*, January 7, E10.

Ross, V. 1995. "Smiths-Coles Book Chain To Be Named Chapters; New Management, Layoffs Coming in Anticipation of Walden-Borders Move into Canada." *Globe and Mail*, April 27, C5.

———. 1999a. "Chapters Venture Sparks Panic. Industry Complains to Ottawa about Retailer's Distribution Company." *Globe and Mail*, May 25, C5.

———. 1999b. "Book Group's Future Uncertain: Attendance Drops at Annual Convention." *Globe and Mail*, June 17, C1.

Scoffield, H. 2000. "Chapters Said to Squeeze Market; Committee Hears Publishers Are Afraid to Speak for Fear of Antagonizing Major Client." *Globe and Mail*, March 3, B4.

———. 2001. "Reisman Finds a Mess at Chapters." *Globe and Mail*, March 14, A1.

Smith, B. 2007. "BookNet Builds Repository of Publishing Industry Data." *ComputerWorld Canada* 23 (13): n.p.

Southworth, N. 1999. "Novel Ideas Needed: Independents Aren't the Target, Chapters CEO Says Wal-Mart, Price/Costco Cited as Real Competition." *Globe and Mail*, June 19, B9.

Stoffman, J. 1999. "Row Rages over Chapters Unit: Book Publishers Balk at Discount Sought by Pegasus." *Toronto Star*, September 15, 1.

Stoffman, J., and J. Ziegler. 2001. "Getting Canada's Book Publishers on Track." *Toronto Star*, June 24, B08.

Strauss, M. 1999. "Chapters Hopes Spinoff Is a Bestseller: Investment in the Retailer's On-line Arm a Boost, but Analysts Warn It Still Has Hurdles to Jump." *Globe and Mail*, June 30, B17.

———. 2000. "Schwartz Suggested Bigger Bid in April: Chapters Circular Reveals $18- to $20-a-share Plan." *Globe and Mail*, December 22, B1.

———. 2002. "Amazon.com Plans Canadian Web Site." *Globe and Mail*, June 1, B1.

————. 2011. "With the Kobo Windfall in Hand, Indigo Chief Executive Heather Reisman Sets Her Sights on Reinventing the Bookseller." *Globe and Mail*, November 12, B1.

Taylor, K. 2011. "CanLit's Latest and Greatest Threat; As the Government Reviews Its Policies, E-commerce Threatens to Upset a Patchwork of Regulations that Has Served the Industry Well." *Globe and Mail*, January 1, R2.

Turner-Riggs. 2007. *The Book Retail Sector in Canada*. Ottawa: Department of Canadian Heritage.

————. 2008. *Book Distribution in Canada's English-language Market*. Ottawa: Department of Canadian Heritage.

Trueheart, C. 1995. "Canadian Booksellers Dread Sheer Volume of U.S. Superstore Invasion." *The Washington Post*, December 13, A33.

Wagman, I. 2010. "On the Policy Reflex in Canadian Communication Studies." *Canadian Journal of Communication* 35 (4): 619–630.

Walker, S. 1995. "Megabookmarts Stake Out Fresh Canadian Turf." *Toronto Star*, June 21, D1.

Williams, L.A. 2011. "Distribution after the Fall of Fenn: Can Distribution Survive in Canada?" *Publishers Weekly*, September 19, S8.

6

Telecommunications: *Plus ça change, plus c'est la même chose?*

Daniel J. Paré

In 2008, Industry Canada auctioned spectrum licenses for advanced wireless services (AWS) and other spectrum in the 2 GHz range. When the auction had ended, Globalive Communications Inc., a new market entrant that had acquired thirty licenses, looked set to establish a national network to compete with Bell, Rogers, and Telus. Some eighteen months later, the company's Wind Mobile service was launched amid a storm of controversy about its eligibility to operate as a telecommunications carrier in Canada. At the core of this dispute lay the highly charged issue of foreign investment in the telecommunications sector. The types of questions raised by the Globalive case are illustrative of a tension that has long been a defining feature of the Canadian communications landscape—namely, the effort to strike a balance between the pursuit of industrial development and other economic objectives on the one hand, and the pursuit of socio-cultural objectives on the other hand (see, for example, Winseck 1998; Babe 1990; Raboy 1990). In the pages that follow, this tension is used as a backdrop for examining recent developments in the Canadian telecommunications environment and for suggesting that little progress has been made in reconciling the differing interpretations of public interest informing Canadian communications policy.

Structure of the Telecommunications Market

The Canadian communications sector is a $57.4 billion industry. It comprises six telecommunications markets (local and access, long distance, Internet, data, private line, and wireless) accounting for $41.7 billion in revenues, and five broadcasting markets (radio, television, broadcast distribution undertakings [BDUs], video on demand [VOD], and pay and pay-per-view [PPV]) accounting for $15.7 billion in revenues (CRTC 2011a). The telecommunications marketplace is dominated by ten facilities-based telecommunications and cable companies (see table 6.1). Within this group, there are four vertically-integrated companies—Bell, Rogers, Shaw, Quebecor/Videotron—that offer services (including service bundles) in each of the above eleven markets. Together, these four companies accounted for almost two-thirds of total communications industry revenues in 2010 (CRTC 2011a; Winseck 2011).

As shown in figure 6.1, wireless services (e.g., mobile telephony, text messaging, roaming, wireless Internet access, and paging services) account for the greatest proportion of revenues ($17.9 billion) generated in the

Table 6.1 Dominant Telecommunications and Cable Companies

Telecommunications Companies	Regional Base	Revenues, 2009 ($ millions)
BCE Inc.	Ontario, Quebec	17,735
TELUS Corporation	Alberta, British Columbia	9,606
Bell Aliant (subsidiary of Bell Canada)	New Brunswick, Newfoundland and Labrador, Nova Scotia, Prince Edward Island	3,174
Manitoba Telecom Services Inc. (MTS–Allstream)	Manitoba	1,810
SaskTel (Saskatchewan Telecommunications Holding Corporation)	Saskatchewan	1,150
Cable Companies		
Rogers Communications Inc.	Ontario	11,731
Shaw Communications Inc.	Alberta, British Columbia, Manitoba, Saskatchewan	3,390
Quebecor/Videotron Ltd.	Quebec	2,000
COGECO Inc.	Ontario, Quebec	1,252
Eastlink	New Brunswick, Newfoundland and Labrador, Nova Scotia, Prince Edward Island	Not available

Source: Adapted from Conference Board of Canada (2011).

telecommunications sector. There are fifty-six licensed cellular operators in the Canadian wireless services marketplace. However, just three firms—Bell, Rogers, and Telus—account for more than 90 per cent of subscriber and revenue market shares (see table 6.2). The remainder of the market is composed of regional carriers, resellers, mobile virtual network operators (MVNOs), and new entrants who acquired spectrum in the 2008 auction. In 2011 both Rogers and Bell began rolling out 4G LTE (long term evolution) networks in Canada's major urban centres. Telus began deploying its 4G LTE network in early 2012.

Figure 6.1 Distribution of 2010 Telecommunications Revenues by Market Sector

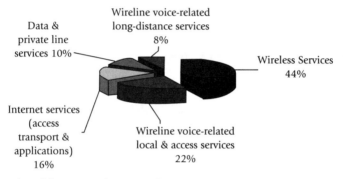

Source: Adapted from CRTC (2011a, 114).

Table 6.2 Subscriber and Revenue Market Share of Canadian Wireless Service Providers, 2010

	Subscriber Market Share (25.8 million subscribers)	Revenue Market Share (Total revenues = $41.7 billion)
Rogers Communications	37%	39%
Bell Canada	29%	28%
TELUS Corporation	27%	27%
Regional Service providers (e.g., MTS-Allstream, SaskTel) and resellers and MVNOs (e.g., Virgin, Koodoo, Fido, and Chatr)	5%	5%
New entrants (e.g., Globalive, Mobilicity, Public Mobile, and Videotron)	2%	1%

Source: Adapted from CRTC (2011a, 160).

While almost the entire population has access to wireless services, in 2010 Canada had the lowest rate of wireless subscribers among the thirty-four OECD member countries—70.66 per 100 inhabitants (see figure 6.2). This relatively low subscription density places Canada 143rd out of 210 countries for which the International Telecommunication Union (ITU) makes data available. Canada also lags far behind its OECD counterparts, and other countries, in the uptake of wireless Internet with only 14.8 active mobile broadband subscribers per 100 inhabitants (ITU 2011a), ranking it 57th in the world.

Figure 6.2 Mobile Cellular Subscribers per 100 Inhabitants, OECD Countries, 2010

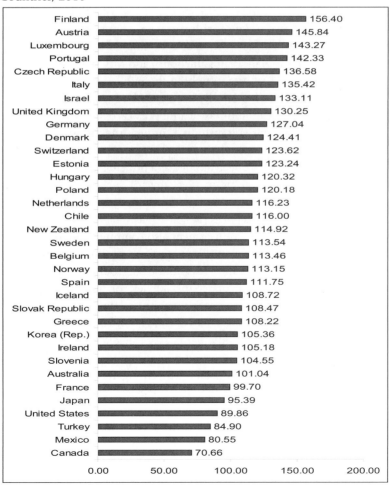

Source: ITU (2011b).

Given the oligopolistic structure of the domestic wireless market, it is not surprising to find that (1) the relatively low levels of wireless uptake coincide with Canadians paying some of the highest prices for mobile services in the world (OECD 2009); and (2) rates of mobile revenue per subscriber are among the highest in the OECD countries (OECD 2011). By the end of 2010, however, the entrance of new players into the domestic wireless marketplace appeared to be contributing to a modest fall in consumer prices (Convergence Consulting Group 2011).

Regulatory Foundations of Telecommunications in Canada

A core task of communications policy is to define and implement governance mechanisms for realizing societal objectives. These objectives tend to be categorized in accordance with various interpretations of the public interest that are themselves rooted in notions of cultural, economic, political, and social welfare (van Cuilenburg and McQuail 2003; Melody 1990). Within Canada, efforts to uphold the public interest in the communications realm have historically been premised upon segregating the provision of communication infrastructure (i.e., carriage) from any determination of what content is transmitted through the infrastructure or of who does the transmitting. Known as the common carrier principle, this regulatory doctrine has long been applied as a principal means of directing activity in telecommunications markets toward such socially desirable ends as interconnectivity among competing networks, provision of universal service, and affordable services (Noam 1994).

The separating of carriage from content also contributed to fostering the industry bifurcation between telephony and broadcasting that remained in place within Canada from the 1920s to the late-1990s. Throughout this period, issues of carriage fell under the purview of telecommunications policy and were concerned foremost with matters relating to technical infrastructure and market conditions. Issues of content, by contrast, were directly associated with broadcasting and other facets of media policy wherein primacy was given to national identity and cultural sovereignty considerations.

There are two key pieces of legislation governing the Canadian telecommunications sector. The first is the *Telecommunications Act, 1993*. It sets out nine objectives for telecommunications policy, including the safeguarding, enriching, and strengthening the social and economic fabric of Canada and its regions; rendering reliable, affordable, high-quality service in urban and rural areas; enhancing efficiency and competitiveness; promoting Canadian ownership and control of telecommunications carriers;

promoting the use of Canadian transmission facilities; fostering increased reliance on market forces and ensuring that regulation is efficient and effective; stimulating telecommunications-related R&D and innovation; responding to the economic and social needs of users; and protecting privacy (see Canada, Minister of Justice 1993, section 7). The second statute is the *Radiocommunications Act, 1985,* which governs the licensing of spectrum and the certification of wireless infrastructure. Its purpose is to ensure the "orderly development and efficient operation of radio communication in Canada" (Canada, Minister of Justice 1985, section 5(1)).

The Canadian Radio-television and Telecommunications Commission (CRTC) is the independent regulatory agency responsible for ensuring that "both the broadcasting and telecommunications systems serve the Canadian public" in accordance with the objectives specified in the *Broadcasting Act* and the *Telecommunications Act* (CRTC 2009a). It reports to Parliament through the Minister of Canadian Heritage.

The federal government has the authority to vary, rescind, or refer back for reconsideration decisions taken by the CRTC as well as the ability to steer the policy agenda through the exercise of certain powers. For example, in 2006, it forcefully nudged the CRTC toward increasingly giving precedence to issues of market conduct as opposed to directing market activity toward specific outcomes by issuing a directive requiring it to "rely on market forces to the maximum extent feasible" and to ensure that the regulatory measures it employs interfere with the "operation of competitive market forces to the minimum extent necessary to meet the policy objectives" (Canada, Minister of Justice 2006, 1).

Liberalization of Canadian Telecommunications

The liberalization of the Canadian telecommunications sector dates back to 1979 when the CRTC ordered the interconnection of CNCP Telecommunications' data and private line network with Bell's local loops. Prior to this decision, customers were required to have separate telephones for each service (see CRTC 1979). Nonetheless, the Canadian telecommunications sector continued to be dominated by regional monopolies throughout the next decade. Indeed, it was not until the early 1990s that the notion of telecommunications being a natural monopoly began to wane.[1] A key turning point in this regard was the introduction, in 1992,

---

1 The historical rationale for characterizing telecommunication as a natural monopoly was linked to two principal considerations: (1) the huge investments required for establishing national telecommunications systems; and (2) concerns that reliance on competitive market forces was

of competition into the market for public long-distance voice services, followed one year later by the enacting of the *Telecommunications Act* which mandated an "increased reliance on market forces" (Canada, Minister of Justice 1993, section 7(f)). These changes where in step with an ascendant information society ideology and the attendant processes of reconvergence that characterized the communications domain throughout the 1990s (see Garnham 2000, 2002; Abramson and Raboy 1999; Winseck 1998).

A defining feature of this period was the notable shift in policy discourse away from viewing communications and information as public goods toward perceiving them as largely technological phenomena that predominantly fell under the auspices of private sector considerations. Accordingly, the regulatory distinction between carriage and content providers was increasingly identified as no longer appropriate to the realities of the emergent global networked economy. Evidence of this can be seen, for example, in the CRTC's 1994 landmark decision introducing a framework for full competition in the telecommunications sector in which it noted that "telecommunications today transcends traditional boundaries and simple definition. It is an industry, a market and a means of doing business that encompasses a constantly evolving range of voice, data and video products and services" (CRTC 1994, n.p.). In 1996, the Canadian government released its convergence policy, *Building the Information Society: Moving Canada into the 21st Century*, which cleared the way for telephone companies and cable service providers to compete in the provision of one another's core services. Shortly thereafter telecommunications companies began to offer television services using satellite and Internet Protocol Television, and the cable companies started competing with wireline telephony by offering voice-over-IP home telephone services.

As the 1990s drew to a close there was growing unease in some quarters that the emphasis given to harnessing the economic and industrial promise of new media technologies was putting Canada's cultural sovereignty at risk. At issue was whether new media services should be treated as carriage or content. After a series of public hearings examining the "rapidly expanding and increasingly available range of communications services collectively referred to as 'New Media'" (CRTC 1998), the CRTC announced that it would "not regulate new media activities on the Internet under the *Broadcasting Act*" because imposing "licensing on new media would not contribute in any

---

unlikely to ensure that Canadians were provided with a reasonable service at reasonable prices. As Melody (2001, 14) writes, the underlying premise was that "a monopoly could achieve maximum economic efficiency by exploiting economies of scale and avoiding duplication of facilities" as well as providing "a single focal point for effective implementation of government policy objectives" such as universal service.

way to its development or to the benefits that it has brought to Canadian users, consumers and businesses" (CRTC 1999a). This was followed by the release of an exemption order, in December 1999, exempting from licensing all new media broadcasting undertakings that provide broadcasting services delivered and accessed over the Internet (CRTC 1999b).

As a result of growing industry consolidation, the relationship between the carriage of communication signals and the content of those signals was revisited on a number of occasions between 2005 and 2007. Of particular concern was the broadcasting of content to mobile devices. The central question was whether content provided via cellular telephones by Bell Mobility Inc., TELUS Mobility, and Rogers Wireless Inc. were "delivered and accessed over the Internet" and, thus, subject to the 1999 exemption. The CRTC ultimately determined that these services did fall within the boundaries of the 1999 exemption (CRTC 2006). This matter was once again on the regulatory agenda throughout 2011 as the CRTC worked to set out a regulatory framework for dealing with vertical integration. The principal issue here was whether the four large vertically-integrated firms operating in the Canadian communication environment—Bell, Rogers, Shaw, and Quebecor—would be able to restrict access to their broadcasting content exclusively to customers of their Internet and wireless services (see CRTC 2011b). While these firms argued that they should have control over their content and platforms, the CRTC ruled that the big four firms could not impede competition in the delivery of wireless services by preventing other wireless service providers (e.g., Telus, SaskTel, MTS–Allstream) from negotiating for the rights to content such as sports programming owned by the big four.

The 2008 Spectrum Auction

In February 2007, Industry Canada initiated a consultation on the auctioning of new mobile spectrum that was designated for "cellular telephony, data, multimedia and IP-based applications, and broadband access, including 3G cellular and other technologies" (Industry Canada 2007a, 1). Six years earlier the three incumbent national carriers—Bell, Rogers, and Telus—had purchased the vast majority of spectrum allotted for personal communication services and then acquired large swaths of fixed wireless spectrum in subsequent auctions. Given this state of affairs, Industry Canada was soliciting comments about the desirability of implementing measures aimed at lowering barriers to entering the wireless marketplace in order to promote facilities-based competition.

It proposed two possible options: (1) setting aside a block or blocks of spectrum; and (2) placing a cap on the amount of spectrum that any one entity could acquire. The former has the advantage of attracting new entrants precisely because they know that they have a real opportunity to at least win what as been set aside. Caps also limit the excessive concentration of spectrum by preventing incumbents from bidding on larger quantities of spectrum than new entrants. Both approaches, however, risk impeding the efficient aggregation of spectrum and facilitating uneconomic entry into the market (Cramton, Kwerel, Rosston, and Skrzypacz 2011).

Industry Canada's consultation generated some sixty comments and replies.[2] With the exception of the incumbent carriers, the comments received were nearly unanimous in their support for introducing measures to facilitate the participation of new entrants. The incumbents maintained that the Canadian wireless market was highly competitive and that the new spectrum should be made available to any firm that was prepared to invest the most in securing and using this resource. Claiming that any favouring of companies in accessing spectrum ran counter to both section 7 of the *Telecommunications Act* and the government's 2006 directive regarding the need for telecommunications policy objectives being achieved through reliance on market forces, they expressed ardent opposition to any interventions in the wireless market.

In late 2007, the Minister of Industry released the policy framework for the forthcoming auction, which stipulated that it would proceed on the basis of dividing the spectrum into eight frequency blocks (covering eight provincial service areas, six large regional service areas, and fifty-nine regional service areas), with three blocks being set aside exclusively for new entrants. In addition, the incumbents were required to make roaming arrangements available to new entrants at "commercial rates that are reasonably comparable to rates that are currently charged to others for similar services" and to allow antenna tower and site sharing (Industry Canada 2007b, 9).

When the auction was completed, fifteen companies had won 282 licenses, earning the government some $4.25 billion for its coffers (see table 6.3). There were twelve new entrants that successfully bid on the spectrum, five of which were regional service providers (Quebecor/Videotron, Shaw, SaskTel, Bragg/Eastlink Communications, and MTS–Allstream) that purchased spectrum to augment their existing product offerings. While the two cable companies in this group—Quebecor/Videotron and Shaw—had each originally planned to use the spectrum they acquired to build wireless networks, Shaw later announced that it would instead use the spectrum

---

2 All of the comments received for this consultation are available at http://www.ic.gc.ca/eic/site/ smt-gst.nsf/eng/sf08769.html. The replies to comments are available at http://www.ic.gc.ca/eic/site/ smt-gst.nsf/eng/sf08785.html.

to expand its broadband Internet and Wi-Fi capacity (CBC News 2011). Another four companies (Blue Canada Wireless, Celluworld Inc., Novus Wireless Inc., and Rich Telecom Corp.) have yet to release details about the service(s) they plan to deploy.

Table 6.3 2008 Mobile AWS Spectrum Auction Winners

License Winners		Cost of Bids	Population Covered
Rogers	59	$999,367,000	30,007,094
TELUS	59	$879,889,000	30,007,094
Bell Mobility Inc.	54	$740,928,000	27,245,106
Globalive Wireless	30	$442,099,000	23,265,134
Bragg Communications	19	$25,628,000	4,886,983
1380057 Alberta Ltd. (Shaw)	18	$189,519,000	9,351,375
9193-2962 Quebec Inc. (Quebecor/Videotron)	17	$554,549,000	14,687,045
Data & Audio-Visual Enterprise Wireless Inc. (Mobilicity)	10	$243,159,000	16,121,864
6934579 Canada Inc. (Public Mobile Inc.)	4	$52,385,077	17,675,254
SaskTel	3	$65,690,000	975,717
6934242 Canada Ltd. (MTS–Allstream)	3	$40,773,750	1,118,283
Novus Wireless Inc.	2	$17,900,000	6,887,060
Rich Telecom Corp.	2	$739,000	133,039
Blue Canada Wireless	1	$1,152,500	1,043,232
Celluworld Inc.	1	$932,000	107,029

Source: Adapted from Industry Canada (2010a).

Of the other three new entrants, one—Public Mobile—serves Toronto and Montreal and intends to expand its network to serve customers throughout the Windsor–Quebec City corridor. The two other new entrants were Mobilicity and Globalive Wireless, both of which are working toward establishing national mobile networks.

Shortly after the 2008 spectrum auction, a controversy erupted regarding Globalive's eligibility to operate as a Canadian telecommunications carrier. The dispute placed the issue of foreign ownership at the forefront of domestic telecommunications policy deliberations. While recognizing that Globalive was owned by Canadians, the CRTC had determined that it was "controlled in fact" by a non-Canadian[3] and, as such, not eligible to operate as a domestic telecommunications carrier (CRTC 2009b, 2009c). Despite being in agreement about Globalive's Canadian ownership status, the federal government and the CRTC differed in their interpretation of who controlled the company, and six weeks later the government issued an order-in-council varying the CRTC's decision thereby permitting Globalive to operate (see Canada, Governor in Council 2009).

Public Mobile (with the support of Telus) then requested a judicial review of this matter. In calling for greater clarity about Canada's foreign investment rules, it argued that the government had exceeded its authority by allowing Globalive to operate in spite of the CRTC's decision, and that it had acted outside of the parameters of the *Telecommunications Act* by restricting its decision solely to Globalive. In February 2010, the Federal Court overturned the government's variance order, citing problems with the legal arguments the government put forward in varying the CRTC's decision (see Berkow 2011). Four months later, Federal Court of Appeals restored, with a unanimous decision, the order-in-council that found Globalive met Canadian ownership and control requirements (see Marlow 2011). In September 2011, Public Mobile launched an appeal motion with the Supreme Court of Canada. The issue remains unresolved at the time of writing.

To Liberalize Foreign Ownership or Not?

In contrast to the broadcast sector, which has been subject to Canadian ownership and control regulations since the late 1950s, foreign owner-ship restrictions are relatively new in the telecommunications sector. In 1984, the Department of Communications granted Rogers Cantel Inc. a license to operate a national cellular network in which it capped foreign ownership at 20 per cent of voting share (Transport Canada 2003). This marked the start of foreign investment restrictions being imposed on telecommunications carriers in Canada. Three years later the Mulroney government released *A Policy Framework for Telecommunications in Canada* which communicated an intention to place statutory limits on the percent-age of allowable foreign ownership for all telecommunications carriers.

---

3 At the time Globalive Wireless was owned by Globalive Investment Holdings Corporation, with two-thirds of the voting shares of this holding company being owned by AAL Holdings Corporation, a Canadian corporation. The remaining one-third of voting shares was owned by Orascom Telecom Holding (Canada) Limited, a subsidiary of Orascom Telecom Holding S.A.E., a holding company controlled by Egyptian billionaire Naguib Sawiris. Orascom financed the vast majority of Globalive's debt and owned 65 per cent of its total equity as well as the Wind and Wind Mobile trademarks. Orascom Telecom has since merged with the Russia-based VimpelCom.

Mirroring existing restrictions in the United States, the objective of the new policy was to "harmonize Canadian policy with that of other countries and ensure our national sovereignty, security and economic, social and cultural well-being" and to protect domestic telecommunications carriers by establishing telecommunications as an exception to the investment regime created within the soon to be finalized Canada–US Free Trade Agreement (Canada, Department of Communications 1987, 1).

With the enacting of the *Telecommunications Act* in 1993, eligibility to operate as a telecommunications common carrier in Canada was made contingent upon the applicant being Canadian-owned and controlled. Section 16(3) of the Act defines Canadian ownership and control as follows:

> (a) not less than eighty per cent of the members of the board of directors of the corporation are individual Canadians; (b) Canadians beneficially own, directly or indirectly, in the aggregate and otherwise than by way of security only, not less than eighty per cent of the corporation's voting shares issued and outstanding; and (c) the corporation is not otherwise controlled by persons that are not Canadians.

The effectuation of these objectives is, in turn, guided by the regulatory framework conferred in *Ownership and Control of Canadian Telecommunication Common Carriers Regulations, 1994*. The regulations contained therein fix the maximum level of foreign ownership of voting shares (including both direct holdings and indirect holdings through a holding company) of a Canadian telecommunications carrier at 46.66 per cent.[4]

Given the unabashed cultural insecurity and economic nationalism informing foreign investment restrictions in the communications domain, successive Canadian governments have avoided engaging directly with this issue since 1993. Today, Canada's foreign investment and ownership policies are among the most restrictive in the OECD, and are well out of step with much of the international community (OECD 2011). These restrictions disproportionally constrain start-up and small player activity by driving up the cost of capital, thereby impeding their ability to pose a viable threat to the large incumbents. By increasing access to capital, the removal

---

4 This figure comprises a 20 per cent direct voting equity in the operating company and a 33.3 per cent voting equity of a holding company that holds the remaining 80 per cent of the operating company.

of foreign investment restrictions should, in theory, foster increased competition and productivity as well as lowering prices.

However, this matter is not so easily resolved precisely because the issue of
foreign investment and ownership pits the objectives of the *Telecommunications
Act* against those of the *Broadcasting Act*. The intersection of these competing
objectives gives rise to a host of socio-economic and socio-cultural concerns.
For instance, might the removal of foreign investment restrictions lead to
Canada's national telecommunications carriers being acquired by foreign
interests? Alternatively, might facilitating access to foreign capital enable
Bell, Rogers, and Telus to further entrench their dominance in the domestic
telecommunications market? While liberalizing ownership restrictions for
broadcasters is not currently on the policy agenda, concerns have also been
expressed about the feasibility of successfully encouraging the production
and availability of Canadian content reflecting Canadian values if the infrastructure for diffusing such content is foreign owned.

Throughout the past decade no less than four government-mandated
studies have been undertaken to investigate these issues. In its 2003
report, the House of Commons Standing Committee on Industry, Science
and Technology concluded that the *Investment Canada Act, 1985*, provided adequate legislative means for ensuring that foreign ownership is
exercised without threatening national sovereignty and in a manner that
is consistent with public interest (see Canada, Standing Committee on
Industry, Science and Technology 2003). It recommended the removal
of all foreign investment restrictions in telecommunications and that
these changes be applied to broadcasting distribution undertakings. The
Telecommunications Policy Review Panel, in 2006, and the Competition
Policy Review Panel, in 2008, called for the adoption of a two-phased
approach to removing foreign ownership restrictions (see Industry
Canada 2006, 2008). Both panels recommended (1) amending the
Telecommunications Act to permit foreign investments in any new telecommunications start-up or any existing telecommunications common carrier
with a market share of 10 per cent or less; and (2) reviewing issues arising
from the separation of policies for the carriage of telecommunications
from broadcasting content policies with a view toward liberalizing foreign
investment rules for both telecommunications and broadcasting.

The most recent of these studies, which was hastened by the 2009
Globalive controversy, is the House of Commons Standing Committee
on Industry, Science and Technology's 2010 report titled *Canada's Foreign
Ownership Rules and Regulations in the Telecommunications Sector*. While the
committee acknowledged that low levels of wireless telephony penetration

combined with the downturn in broadband penetration "are symptomatic that all is not well in Canada's telecommunications industry," this committee's enthusiasm for removing foreign investment barriers was far more tempered than that expressed by the three previous panels (see Canada, Standing Committee on Industry, Science and Technology 2010). Its principal recommendation focused on the need for greater clarity in the metric used to establish whether a company is controlled "in fact" by Canadians in determinations of eligibility to operate as a Canadian telecommunications carrier.

Coinciding with the publication of the latter report, Industry Canada released a consultation document in which it proposed three options for altering existing foreign investment restrictions for the telecommunications sector (Industry Canada 2010b). They were:

1) symmetrical liberalization of telecommunications and broadcasting by increasing the limit of allowable foreign investment to 49 per cent;
2) removal of foreign investment restrictions on telecommunications carriers that have less than 10 per cent of revenue market share;
3) complete removal of all foreign investment restrictions in the telecommunications sector.

The comments received in response to these proposals can be divided into four categories.[5] In rejecting outright options two and three,[6] the three incumbents expressed concerns about the potential for limiting the liberalization of foreign investment restrictions to the telecommunications sector to unduly benefit "pure players" at the expense of firms that are subject to both the *Telecommunications Act* and the *Broadcasting Act*. There was no one option predominantly favoured by smaller telecommunications carriers and cable companies. Indeed, a number of these firms argued that the three policy options were inadequate. Here, too, the potential for the proposed options to disadvantage firms with integrated activities in both the telecommunications and broadcasting sectors was a source of apprehension. By contrast, Globalive, Mobilicity,

---

5 All of the comments received for this consultation are available at http://www.ic.gc.ca/eic/site/smt-gst.nsf/eng/sf09919.html.
6 Both Bell and Rogers expressed support for option one, while Telus rejected all three options. Instead, it called for the complete removal of all foreign ownership restrictions in both the telecommunications and broadcasting sectors to be combined with the maintenance of restrictions on ownership of broadcast channels and specialty services, and the adoption of safeguards with respect to the vertical integration of broadcast carriers.

and Public Mobile all expressed resounding support for option two. Among the submissions from organizations representing various facets of the cultural industries there was near unanimous support for the maintenance of the status quo.

At the time of writing, there is another spectrum auction, in the 700 MHz band, planned for late 2012. Spectrum in this frequency band is extremely valuable because it covers long distances and is particularly useful for expanding the rural reach of mobile broadband services. The key objective this time around is to promote greater competition in the wireless broadband market (Industry Canada 2010c). Not surprisingly, the positions of the three incumbents regarding the use of set-asides and/or spectrum caps for the upcoming auction are little changed from those advanced in 2008.[7] Likewise, the majority of other actors, including the new entrants who acquired spectrum in 2008, continue to express support for the implementation intervention measures akin to those employed in 2008. However, given the regulatory challenges with which it has had to contend since winning its spectrum licenses, Globalive has indicated that it likely will not participate in the 2012 auction unless the issue of foreign investment is resolved (Trichur, Chase, and Marlow 2011).

Back to the Future?

More than twenty years ago, Babe (1990) argued that the history of tele-communications in Canada was marked by fluctuations between periods of competition and monopoly. The past few years have witnessed the Canadian government seeking to foster greater facilities-based competition in the wireless sector. The issue today is whether the independent new entrants who acquired spectrum in 2008—and those who will acquire spectrum in the near future—will be able to become viable competitors to Bell, Rogers, and Telus or whether this period will become just another historical oscillation. The manner in which the issue of foreign investment and ownership is resolved will play a pivotal role in the outcome.

Proponents of easing foreign investment restrictions tend to privilege commercial- and consumer-oriented considerations (e.g., increased competition, improved services, greater innovation, lower prices, harmonizing Canadian policy with that of Canada's trading partners) and question the efficacy of imposing restrictions on externally based investments as a means for protecting national sovereignty and cultural well-being. This

---

7 All the comments received from the public consultation pertaining to this auction are available at http://www.ic.gc.ca/eic/site/smt-gst.nsf/eng/sf09997.html.

view assumes that liberalizing foreign investment restrictions will likely increase competition and productivity by eroding the oligopolistic structure of the Canadian telecommunications market and directly benefit consumers. Here, we observe an interpretation of public interest that is closely tied to conceptions of economic welfare and utility. By contrast, advocates for retaining the status quo tend to advance interpretations of the public interest that privilege socio-cultural considerations and which maintain that the carriage of communication signals cannot be separated from cultural identity, national sovereignty, and a democratic public sphere. They portend a domino effect in which the removal of foreign investment restrictions on telecommunications leads to pressure to accord broadcasting distribution industries the same treatment as telecommunications carriers which, in turn, inevitably affects broadcasting content, thereby irreparably harming Canadian cultural sovereignty.

A fundamental question in the current telecommunications environment is whether the hypothesized benefits of easing and/or eliminating foreign investment and ownership restrictions can be expected to materialize in a manner that strikes an appropriate balance between the various welfare considerations that constitute the public interest. The House of Commons Standing Committee on Industry, Science and Technology offers a particularly lucid articulation of the Gordian Knot with which Canadian policy-makers must now contend in this regard:

> While the removal of foreign ownership restrictions for telecommunications common carriers and broadcasting distributions undertakings may improve competition, services and prices, the following question also needs an urgent response: How can Canada's cultural sovereignty, with respect to broadcasting content, be best protected and promoted in an era of potential complete convergence between television and Internet? (Canada Standing Committee on Industry, Science and Technology 2010, 44)

While the underlying technologies have changed, the committee's observation directly corresponds with the types of questions about the relationship between telecommunications and society that the Telecommission sought to address some forty years ago (see Canada, Department of Communications 1971). The central enigma then, as now, centres on how to reconcile economic and socio-cultural objectives in view of technological innovations. Seen in this light, contemporary debate about foreign investment in telecommunications really is not all that new. It simply is the latest *cause célèbre* for articulating contending visions of the public interest that have resonated throughout the history of Canadian communications policy.

References

Abramson, Bram Dov, and Marc Raboy. 1999. "Policy Globalization and the 'Information society': A View from Canada." *Telecommunications Policy*, 23 (10-11): 775–791.

Babe, Robert E. 1990. *Telecommunications in Canada*. Toronto: University of Toronto Press.

Berkow, Jameson. 2011. "Globalive Joins Ottawa in Appealing Federal Court Ruling." *Financial Post*, February 17. Accessed September 30, 2011. http://business.financialpost.com/2011/02/17/globalive-joins-ottawa-in-appealing-federal-court-ruling.

Canada. Department of Communications. 1971. *Instant World: A Report on Telecommunications in Canada*. Ottawa: Information Canada.

———. 1987. *Highlights of the Telecommunications Policy for Canada*. Ottawa: Department of Communications.

Canada. Governor in Council. 2009. *Order-in-Council PC 2009-2008*. Accessed September 30, 2011. http://www.ic.gc.ca/eic/site/ic1.nsf/vwapj/PC2009-2008-eng.pdf/$file/PC2009-2008-eng.pdf.

Canada. Minister of Justice. 1985. *Radiocommunication Act*. C. R-2. Accessed September 30, 2011. http://laws-lois.justice.gc.ca/PDF/R-2.pdf.

———. 1991. *Broadcasting Act*. C. 11. Accessed September 30, 2011. http://laws.justice.gc.ca/PDF/B-9.01.pdf.

———. 1993. *Telecommunications Act*. C. 38. Accessed September 30, 2011. http://laws-lois.justice.gc.ca/PDF/T-3.4.pdf.

———. 1994. *Ownership and Control of Canadian Telecommunication Common Carriers Regulations*. SOR/94-667. Accessed September 30, 2011. http://laws.justice.gc.ca/PDF/SOR-94-667.pdf.

———. 2006. *Order Issuing a Direction to the CRTC on Implementing the Canadian Telecommunications Policy Objectives SOR/2006-355*. Accessed September 30, 2011. http://laws-lois.justice.gc.ca/PDF/SOR-2006-355.pdf.

Canada. Standing Committee on Industry, Science and Technology. 2003. *Opening Canadian Communications to the World*. Ottawa: Communication Canada. Accessed September 30, 2011. http://www.parl.gc.ca/content/hoc/Committee/372/INST/Reports/RP1032302/instrp03/instrp03-e.pdf.

———. 2010. *Canada's Foreign Ownership Rules and Regulations in the Telecommunications Sector*. Ottawa: Public Works and Government Services Canada. Accessed September 30, 2011. http://www.parl.gc.ca/content/hoc/Committee/403/INDU/Reports/RP4618793/indurp05/indurp05-e.pdf.

CRTC (Canadian Radio-television and Telecommunications Commission). 1979. *Telecom Decision CRTC 79-11. CNCP Telecommunications: Interconnection with Bell Canada* (May 17).

———. 1994. *Telecom Decision CRTC 94-19* (September 16). Accessed September 30, 2011. http://crtc.gc.ca/eng/archive/1994%5CDT94-19.htm.

———. 1998. *Telecom Public Notice CRTC 98-20* (July 31). Accessed September 30, 2011. http://crtc.gc.ca/eng/archive/1998/PT98-20.htm.

———. 1999a. *Telecom Public Notice CRTC 99-14* (May 17). Accessed September 30, 2011. http://www.crtc.gc.ca/eng/archive/1999/pt99-14.htm.

———. 1999b. *Public Notice CRTC 1999-197, Exemption Order for New Media Broadcasting Undertakings* (December 17). Accessed September 30, 2011. http://www.crtc.gc.ca/eng/archive/1999/pb99-197.htm.

———. 2006. *Broadcasting Notice CRTC 2006-47, Regulatory Framework for Mobile Television Broadcasting Services* (April 12). Accessed September 30, 2011. http://www.crtc.gc.ca/eng/archive/2006/pb2006-47.pdf.

———. 2009a. "About the CRTC." Accessed September 30, 2011. http://www.crtc.gc.ca/eng/backgrnd/brochures/b29903.htm.

———. 2009b. *Telecom Decision CRTC 2009-678, Review of Globalive Wireless Management Corp. under the Canadian Ownership and Control Regime* (October 29). Accessed September 30, 2011. http://

www.crtc.gc.ca/eng/archive/2009/2009-678.htm.
——. 2009c. *Telecom Decision CRTC 2009-678-1, Review of Globalive Wireless Management Corp. under the Canadian Ownership and Control Regime* (November 4). Accessed September 30, 2011. http://www.crtc.gc.ca/eng/archive/2009/2009-678-1.htm.
——. 2011a. *Communications Monitoring Report (July). Accessed* September 30, 2011. *http://www. crtc.gc.ca/eng/publications/reports/policymonitoring/2011/cmr2011.pdf.*
——. 2011b. *Broadcasting Regulatory Policy CRTC 2011-601: Regulatory Framework Relating to Vertical Integration* (September 21). Accessed September 30, 2011. http://www.crtc.gc.ca/eng/archive/2011/2011-601.pdf?
CBC News. 2011. "Shaw Cancels Cellular Network Plans." Accessed September 30, 2011. http://www.cbc.ca/news/technology/story/2011/09/01/shaw-wireless.html.
Conference Board of Canada. 2011. *Canadian Industrial Outlook, Spring: Canada's Telecommunications Industry.* Ottawa: Conference Board of Canada.
Convergence Consulting Group. 2011. *Canadian Wireless 2009-2014: Assessing the Impact of New Entrants, September 2011.* Accessed September 30, 2011. http://www.convergenceonline.com/downloads/CanWireless2011.pdf.
Cramton, Peter, Evan Kwerel, Gregory Rosston, and Andrzej Skrzypacz. 2011. *Using Spectrum Auctions to Enhance Competition in Wireless Services.* SIEPR Discussion Paper No. 10-015. Stanford Institute for Economic Policy Research Stanford University. Accessed September 30, 2011. http://www-siepr.stanford.edu/repec/sip/10-015.pdf.
Garnham, Nicholas. 2000. "'Information Society' as Theory or Ideology: A Critical Perspective on Technology, Education, and Employment in the Information Age. *Information, Communication, and Society* 3 (2): 139–152.
——. 2002. "Information Society: Myth or Reality." Paper presented at 2001 Bugs: Globalism and Pluralism, Montreal, April 24–27. Accessed September 30, 2011. http://www.er.uqam.ca/nobel/gricis/actes/bogues/Garnham.pdf.
Industry Canada. 1996. *Building the Information Society: Moving Canada into the 21st Century.* Ottawa: Information Highway Advisory Council Secretariat
——. 2006. *Telecommunications Policy Review Panel: Final Report.* Ottawa: Public Works and Government Services Canada. Accessed September 30, 2011. http://www.ic.gc.ca/eic/site/smt-gst.nsf/vwapj/tprp-final-report-2006.pdf/$FILE/tprp-final-report-2006.pdf.
——. 2007a. *Notice No. DGTP-002-07 – Consultation on a Framework to Auction Spectrum in the 2 GHz Range including Advanced Wireless Services.* Accessed September 30, 2011. http://www.ic.gc.ca/eic/site/smt-gst.nsf/vwapj/aws-consultation-e.pdf/$FILE/aws-consultation-e.pdf.
——. 2007b. *Policy Framework for the Auction for Spectrum Licences for Advanced Wireless Services and other Spectrum in the 2 GHz Range.* Accessed September 30, 2011. http://www.ic.gc.ca/eic/site/smt-gst.nsf/vwapj/awspolicy-e.pdf/$FILE/awspolicy-e.pdf.
——. 2008. *Compete to Win: Final Report.* Competition Policy Review Panel. Ottawa: Public Works and Government Services Canada. Accessed September 30, 2011. http://www.ic.gc.ca/eic/site/cprp-gepmc.nsf/vwapj/Compete_to_Win.pdf/$FILE/Compete_to_Win.pdf.
———. 2010a. "Auction of Spectrum Licences for Advanced Wireless Services and Other Spectrum in the 2 GHz Range: Licence Winners." Accessed September 30, 2011. http://www.ic.gc.ca/eic/site/smt-gst.nsf/eng/sf09002.html.
——. 2010b. *Opening Canada's Doors to Foreign Investment in Telecommunications: Options for Reform.* Accessed September 30, 2011. http://www.ic.gc.ca/eic/site/smt-gst.nsf/vwapj/TelecomInvestment-eng.pdf/$file/TelecomInvestment-eng.pdf.
——. 2010c. *Notice No. SMSE-018-10: Consultation on a Policy and Technical Framework for the 700 MHz Band and Aspects Related to Commercial Mobile Spectrum* (November 30). Accessed September 30, 2011. http://www.ic.gc.ca/eic/site/smt-gst.nsf/vwapj/smse018e.pdf/$FILE/smse018e.pdf.
ITU (International Telecommunications Union). 2011a. *Measuring the Information Society.* Geneva: International Telecommunications Union. Accessed September 30, 2011. http://www.itu.int/ITU-D/ict/publications/idi/2011/Material/MIS_2011_without_annex_5.pdf.
——. 2011b. "Website Database." Accessed September 30, 2011. http://www.itu.int/ITU-D/ict/statistics.
Marlow, Iain. 2011. "Public Mobile Seeks to Appeal Globalive Ruling to Top Court." *Globe and Mail,* June 9. Accessed September 30, 2011. http://www.theglobeandmail.com/report-on-business/public-mobile-seeks-to-appeal-globalive-ruling-to-top-court/article2054509.

Melody, William H. 1990. "Communication Policy in the Global Information Economy: Wither the Public Interest?" In *Public Communication: The New Imperatives*, ed. Marjorie Ferguson, 16-39. London: Sage.

———. 2001. "Policy Objectives and Models of Regulation." In *Telecom Reform: Principles, Policies and Regulatory Practice*, ed.William H. Melody, 11-24. Lyngby, DTU. Accessed September 30, 2011. http://lirne.net/test/wp-content/uploads/2007/02/telecomreform.pdf.

Noam, Eli. 1994. *"Beyond Liberalization II: The Impending Doom of Common Carriage." Telecommunications Policy* 18 (4): 286–294.

OECD. 2009. *OECD Communications Outlook 2009*. Paris: OECD Publishing.

———. 2011. *OECD Communications Outlook 2011*. Paris: OECD Publishing

Raboy, Marc. 1990. *Missed Opportunities: The Story behind Canada's Broadcasting Policy*. Montreal: McGill-Queen's University Press.

———. 1996. "Cultural Sovereignty, Public Participation and Democratization of the Public Sphere: The Canadian Debate on the New Information Infrastructure." *Communications and Strategies 21*: 51–76.

Transport Canada. 2003. *Restrictions on Foreign Ownership in Canada – TP 14500E*. Ottawa: Policy Research Branch Strategic Policy Directorate Policy Group. Accessed September 30, 2011 http://www.tc.gc.ca/media/documents/policy/tp14500e.pdf.

Trichur, Rita, Steven Chase, and Iain Marlow. 2011. "Loosen Grip of Big Three, Telco Upstarts Tell Ottawa." *Globe and Mail*, November 23. Accessed September 30, 2011. http://www.theglobeandmail.com/report-on-business/loosen-grip-of-big-three-telco-upstarts-tell-ottawa/article2246926.

van Cuilenburg, Jan, and Denis McQuail. 2003. "Media Policy Paradigm Shifts: Towards a New Communications Policy Paradigm." *European Journal of Communication* 18 (2): 181–207.

Winseck, Dwayne. 1998. *Reconvergence: A Political Economy of Telecommunications in Canada*. Cresskill: Hampton Press.

———. 2011. "Big Media in the Hot Seat at CRTC Hearings." *Globe and Mail*, June 20. Accessed September 30, 2011. http://www.theglobeandmail.com/news/technology/digital-culture/dwayne-winseck/big-media-in-the-hot-seat-at-crtc-hearings/article2067311.

PART TWO
CULTURAL INDUSTRIES: DISCOURSE, METHODS, POLICY

7

Continuity and Change in the Discourse of Canada's Cultural Industries

Zoë Druick

> *There is clear empirical evidence that Canada's culture sector, encompassing arts and culture industries, generates a wealth of contributions to Canada's cultural, social, and economic fabric.*
> —The Conference Board of Canada, *Valuing Culture: Measuring and Understanding Canada's Creative Economy*

The equation apparent in the above quotation between Canadian culture and the industrial production of wealth and social benefits has become normative and ubiquitous. According to this logic, the production of art and culture can and must be seen in terms of what economists call its external benefits. Yet, upon reflection, it makes numerous assumptions about the role of culture in both the economy and the national imaginary. And, although it is possible to research the creative economy and the cultural industries starting from this unquestioned presupposition, it might make more sense for researchers to begin by asking themselves some preliminary questions. For instance, when I set out to study the cultural industries, what am I hoping to study and why? One might go further to query: Do the cultural industries actually exist as an object of study before I name them as such? And, presuming that they do, what—exactly—makes them distinctively Canadian? The hope for a clear methodological approach to the study of the object "the cultural industries in Canada" is perhaps not as well founded as we might assume. It is based, rather, on numerous assumptions and frameworks that ought to be fully examined by the researcher. Through historical and discursive analysis, this chapter aims to expose and destabilize some of the assumptions that have accrued around national culture and creative or cultural industries in order to leave researchers better prepared to engage in projects that study the "cultural industries in Canada." I hope to show that many of the associations we have come to take for granted, including the relationships between the parts of the term "cultural industries," have taken shape over time in relation to numerous economic and discursive struggles around common-sense notions of the nature and value of creativity. In addition, potentially unresolvable contradictions and tensions within government and industry about cultural objectives have meant that Canada has ended up with sophisticated communications policy yet underdeveloped cultural industries, a challenge for any researcher.

When studying the cultural industries, perhaps our first question ought to be: What is culture? One widespread definition sees culture as referring both to a whole way of life and to the products of the culture that expresses them (McAnany 1986, 14). The "whole way of life"—or anthropologic-al—concept of culture refers to all meaning-making, symbol- and text-producing enterprises, many of which have become central to a digital economy based on information, knowledge, and creation. What, then, are the cultural industries? David Hesmondhalgh (2007) argues that these industries are connected through their overlapping labour pools, their appeal to the same audiences, their ownership profiles, and their regula-tion by a set of interlinked policies (13). The cultural industries are thus defined both by the type of product they make and the infrastructure and political economy that underpin them. The exclusions are just as import-ant as the inclusions: media industries, especially those concerned with audio-visual production, are always included in the category of cultural industries, while clothing, food, sports, gardening, and other meaning-bearing activities—many of which also occur at an industrial scale—usu-ally are not. High arts, often considered part of the creative economy, are not usually included in the category of cultural industries, indicating that a divide between high and low culture is implied in the term, a point to which we shall return. In short, the term "cultural industries" refers in the main to the production scale and conventions associated with the global, corporate audio-visual media, a field overwhelmingly dominated in the English-speaking world by Hollywood. The word "culture" in the term "cultural industry" thus maps but imperfectly onto an idea of a "whole way of life" in Canada. We are speaking then of culture that, in the main, expresses and supports industrial values and concerns.

A series of shifts and intensifications that have taken place since the late

1970s have made the cultural industries even more central to social, political, and economic life. For instance, changes to industrial policies have led to the increased globalization of capital as well as drastic changes to the approach by contemporary nation-states to considerations of welfare and the public good that are often referred to in shorthand as neoliberalization. According to David Harvey (2005), "Neoliberalism is in the first instance a theory of political economic practices that proposes that human well-being can best be advanced by liberating individual entrepreneurial freedoms and skills within an institutional framework characterized by strong private property rights, free markets, and free trade" (2). As Gattinger and Saint-Pierre (2010) note, the "neoliberal turn" has meant that the concepts of culture, cultural policy, and economy have come to be linked; that economic imperatives "come to dominate cultural policy rationales, objectives, and targets"; that government intervention has been "reduced or reoriented"; and that cultural responsibilities are "decentralized or devolved to lower levels of government or to non-government actors" (280). These rationales have taken the place of previous discourses about the contribution of arts and culture on their own terms and of cultural objectives as an end in themselves (*ibid.*, 281). Within this framework, which has ushered in the digital economy, culture has arguably taken on increased significance in public discourse as the new paradigm of autonomous, flexible, creative labour, as the primary activity in an economy based on symbol production, and as the core site of innovation in the knowledge economy (Deuze 2007). As French sociologist Bernard Miège (2011) observes, for better or worse, the cultural industries are "now seen to be particularly representative and emblematic of the general transformations of contemporary capitalism and globalization" (90–91).

Inroads have certainly been made into Canadian social and political life by neoliberal ideas about the reversal of Keynesian welfare economics and the transformation of the state into an enabler of markets to meet needs formerly administered by government policy and programs; however, it is important to note that these ideas often exist in uneven and hybrid ways (Gattinger and Saint-Pierre 2010; Jeannotte 2010; Marontate and Murray 2010). In addition, the global economic crash of 2008 demonstrated the culmination—and in many ways the failure—of thirty years of intensive neoliberalization in the global North. The lopsided struggles to move social and economic policies in that general direction are thus ongoing, complex, and unresolved. Due to their power as shapers of cultural perspectives, when engaged in research on the cultural industries, one must be particularly vigilant about the processes of neoliberalization and how they are framed.

Common use of terms such as "creative economy," "knowledge economy," "information economy," "digital economy," and the "cultural industries" makes us aware of the instrumental view of culture that has come to prevail in the current economic configuration. These terms all present culture as a tool of economic expansion while also clearly considering it to be something special and apart from economic considerations (or else there would be no need to add a modifier to the word "industry"). The paradox of tying culture to economy must be kept central in any analysis of the concept. As we shall see, the initial impulse for discussing the cultural industries was to draw attention to the contradictions of culture's relationship to capital—a project with continued value as increasingly, with the rise of cultural economics since the 1980s, the terms are tied together without any critical intent or irony.

In the sections that follow, I consider the history of the terms and how they have been mobilized in the Canadian context. I map the emergence of neoliberal and governmental logics in the administration and framing of culture and the different strands of scholarly and policy discourse that have helped to shape the key debates of the 1980s and 1990s. I conclude by offering a synthesis of the ideas covered in the chapter.

Connection of the Culture Industries to Both "Culture" and "Industry"

Given the presumed connection today between culture and capitalism, it is perhaps surprising to remember that the terms "Culture Industry" and "the cultural industries" both began as critical ones. I will consider them each in turn. Before the publication of *The Dialectic of Enlightenment* (1944) by Max Horkheimer and Theodor Adorno, two researchers from the Frankfurt School who were in exile in the United States during the Second World War, what we might now call the Culture Industry was usually referred to as "mass culture." Mass culture traces its roots to the nineteenth century and the proliferation of cheap mass-market cultural goods, such as magazines and low-brow literature. Mass culture was presumed by critics to reside at the opposite end of the spectrum from high culture, which consisted of legitimate forms of culture enshrined in institutions such as museums and opera houses, and the terms had a set of gendered associations, with mass culture considered to be both feminized and essentially worthless (Huyssen 1986).

Exposing the lower classes to high culture was seen to be a way of inculcating civilized values, an idea that can be read in Matthew Arnold's educational classic *Culture and Anarchy* (1882) (see also Bennett 1998).

Yet, as Horkheimer and Adorno trace in the *Dialectic of Enlightenment*, the shadow of enlightenment humanism of this kind is technological rationality, industrial logic writ large. According to them, the role of culture is not to compensate for instrumental and dehumanizing systems of modernity, but to reflect this tension between opposing tendencies.

This analysis becomes especially important in their chapter on the Culture Industry, "Enlightenment as Mass Deception." Contrary to common caricature, the German theorists did not dismiss mass culture out of hand so much as draw attention to the fact that in Hollywood films, art has been fully subordinated to industrial concerns.

> The people at the top are no longer so interested in concealing monopoly: as its violence becomes more open, so its power grows. Movies and radio need no longer pretend to be art. The truth that they are just business is made into an ideology in order to justify the rubbish they deliberately produce. They call themselves industries; and when their directors' incomes are published, any doubt about the social utility of the finished products is removed. (Horkheimer and Adorno 1944, 121)

Within the Culture Industry, advertising, the bald-faced intention to sell, is merged with cultural products both technically and economically at an unprecedented scale (*ibid.*, 163). In a follow-up essay written close to two decades later, Adorno clarified that the term "Culture Industry" was chosen to create differentiation from popular culture; it was not to be confused with a culture of the people, although masses were its target market. They were "commodities through and through" (Adorno 1967, 86). In other writing, Adorno (1967) shows that he is aware that all culture, high and low alike, is reliant upon the instrumental logic of administration, an observation that has a good deal of resonance for the study of cultural policy today. The ideas of the Frankfurt School are still important for their articulation of a critique of mainstream commercial culture that implied a connection between how a cultural product was made and what it was about. In addition, they drew together disparate aspects of commercially produced culture in order to create a new object of analysis: The Culture Industry. Their ideas require some updating to keep abreast of the changes to production and consumption undertaken by the cultural industries in recent decades, but the kernel of insight at the heart of their observations need not be lost.

There was a Marxist revival of the cultural industries discourse in France around the time of the student and worker uprisings in May 1968. Although Adorno and Horkheimer were read with interest at that time, there was a sense that the Culture Industry discourse didn't allow for the possibilities of new art forms. As Bernard Miège (1989) puts it, "If technologies unquestionably accompany the development of cultural commodities, they also open up new directions in art. The refusal of commoditization mustn't bring in its wake a distrust for technology and artistic innovation" (10). The pluralization of the term "cultural industries" was, then, a conscious attempt to broaden out the analysis by placing more nuance on the object being studied. Increased differentiation between cultural industries, some of which were dominated by American products (such as theatrical film) and others (such as music) which were not, would allow for a more incisive set of critiques. In other words, different logics were seen to regulate each industry as they became more developed and differentiated.

As capitalism began to be reconfigured on a global scale in the 1960s and 1970s, bringing about the beginning of deindustrialization in the West, two themes in investigations into the cultural industries were merged: the "fate of art and aesthetic creativity" on one hand and "the domination of the economy and its consequences for democratic culture" on the other (Miège 2011, 84). The relationship of the marketplace and industrial concerns to both creativity and democracy became a central concern. Was freedom of expression even a possibility in this new corporate-controlled and technology-mediated public sphere? Connections being made between art and democracy in real rather than formal terms also gave birth in the early 1970s to the reform agenda of the field of critical political economy of communication. As Jim McGuigan (2003) characterizes this position, "Without provision for public accountability and control, and without some attempt to foster alternatives to an overbearing state and untrammelled market forces, the prospects for democratic debate and cultural experiment would be curtailed and, indeed, imperilled" (35).

While mid-century policy initiatives in the field of the arts attempted to democratize culture by making high culture in museums and symphony halls more accessible, there was a backlash against the presumed value of high art itself. The valuing of popular culture seemed a more likely route to cultural democracy than the high-handed strategy of trying to influence tastes. Yet this shift in theorizing about the cultural industries and cultural policy as a site of possible democratization had an unforeseen backlash. When left-wing and Labour governments in Western democracies came to power in the 1980s, they enlisted critical scholars to help them forge new, more inclusive cultural policy. For instance, British communications scholar Nicholas Garnham was hired by the Greater London Council to "help craft a 'cultural industries' strategy that would remove class distinctions from cultural policy and expand the state's view of culture to encompass not only such things as art museums, literature, and opera but television, radio, and pop music" (Lewis and Miller 2003, 7). As Jim McGuigan (2003) points out, the outcome of this convergence of cultural reformers and state agencies was for "'culture' to be resituated within the economistic and technicist discourses of public policy and in this way… tied into the governmentality of communications media on industrial and economic grounds" (36). Significantly, it was within this conjuncture that New Labour pioneered the deployment of the term "creative industries" as well (Miège 2011, 103).

In Australia, between 1983 and 1996, Labour governments enlisted academics to help solve social and cultural problems. Their blueprint for cultural development, called "Platform," aimed to establish Australia as a "creative nation." On the one hand, this was based on a progressive vision of post-industrial society; on the other, it ushered in the neoliberalization of the cultural sector and the emergence of the digital economy.[1] Paradoxically, then, even as the culture industry critique became mainstream (it could be found throughout UNESCO documents, for instance), culture itself became increasingly instrumentalized as it was diversified through a variety of social and economic locations. The struggle for the democratization of cultural access that had preoccupied Cold War cultural politics morphed into the quest for cultural democracy, the valuing of different cultural tastes—something that might be left to the market.

In the 1980s, partly as a result of critiques brought by scholars of the elitism of cultural policy in educational/cultural institutions, the discourse shifted from one of cultural quality to one of diversity and multiplicity. In the UK, for instance, New Labour seized on this expansion of the notion of culture by broadening ministerial responsibility for the arts to include the electronic media and other popular-cultural industries in the late 1990s (Lewis and Miller 2003, 7). Was a check required on the cultural industries, or would the market provide the requisite diversity? Arguably less governmental, the affirmation of cultural difference and the attention to the implicit norms of class inherent in cultural development converged

---

1 It is interesting to note that a robust discourse of cultural policy does not exist in the US. Jonathan Sterne has hypothesized that this may be because, in the US, communication policy is discussed as communication law. See Sterne (2002, 62).

with the push by cultural industries themselves for more products to be made available to the market. In this way, an emphasis on the need for cultural products reflecting multiple class, gender, and cultural perspectives and realities dovetailed with commodity feminism and market multiculturalism, with audiences subdivided by taste into smaller and more desirable groups of affluent consumers with cultural differences (Canclini 2005; Gill 2007). As we will see, a similar shift occurred in the Canadian case, ushering in the political changes that would lead to the signing of the Free Trade Agreement.

Dependency and Governmentality: The Canadian Case

During the postwar decades, in a quest for the democratization of culture—a term that was tantamount to broadening public access—criticisms of mass culture became commonplace. Some, like French sociologist Augustin Girard ([1972] 1983) of the French Ministry of Cultural Affairs, argued that the scale made possible by cultural industries should be marshalled by high cultural forms, such as theatre and opera, thereby creating the democratic access to culture that cultural institutions were unable to deliver. In Canada, a strong tendency since the Massey Commission had been the hope that electronic media might promote high culture in precisely this way. But audiences were not always interested. Time and again, Canadians remained "absent" from the cultural experiences deemed best for them (Acland 1997).

The Royal Commission on the Development of the Arts, Letters and Sciences (The Massey Commission), which convened from 1949 to 1951, was obviously aware of criticisms of American mass culture, if not of the Frankfurt School. Although its title seemed to point at high arts, one of the goals of the commission was clearly to put a Canadian signature on twentieth-century electronic media. Another aim was to put policy rationales behind the establishment of publicly funded cultural institutions in Canada. This marked an important shift from small-scale, artisanal, community-level cultural production to culture with more national (and international) ambition (Tippett 1990; Druick 2006). However, the Massey moment—with its emphasis on quality and its link to the foundation of CBC TV, the Canada Council for the Arts, and the National Library—has also served to suppress histories about more industrial forms of culture made in Canada, such as the proliferation of private radio stations and companies dedicated to the production of industrial film, which have subsequently been marginalized from Canadian cultural history (Canadian

Educational, Sponsored and Industrial Film Archive n.d.; Wagman 2010). In this case, the determination by critical communication scholars that nothing outside of anti-American public media infrastructure was worthy of attention meant that the object of the Canadian cultural industries was rendered invisible. In short, cultural critique might have been the purview of Canadians, but the cultural industries were resolutely American.

Indeed, Gattinger and Saint-Pierre (2010) argue that as far as provincial governments were concerned, the dominant objectives for culture in English Canada before the 1970s were access and education (289). Outside of Quebec, governments were reluctant to become involved with patronage for the arts. At the national level, however, American media were seen to threaten the possibility of creating a meaningful and distinct national identity and the Canadian state was obliged to wade into the fray, especially after Quebec began an aggressive policy of cultural development in the 1960s (*ibid.*, 283). In the 1970s, cultural nationalism converged with political economic critiques of American media dominance to entrench a vision of state-supported, distinctive, and meritorious national culture. "Culture," writes Ryan Edwardson (2008), "offered a new intelligentsia a means by which to reclaim Canadian sovereignty from forces of imperialism" (159). The book and magazine industries in particular were reinforced by the commitment to national literary culture (Wright 2001).

Yet Canadian content regulations, first adopted for television in 1961, are emblematic of the struggles that were being waged around the significance of Canadian cultural industries. On one hand, the quantification of Canadian content had the effect of ensuring that more of it was produced and aired. On the other, it couldn't account for the quality of the content, which was intangible and unmeasurable. Thus broadcast quotas demonstrated that Canadian culture couldn't adhere to a supply-side model. At the same time, the new nationalism emerging in the 1960s and 1970s committed itself to creating distinctive Canadian fare. In theory at least, this content could fill the space that the industry was currently mandated to allot to it—so long as advertisers were amenable. In effect, notions about Canadian cultural production inherently challenging American hegemony were often underwritten by economic imperatives for Canadian media companies.

This presumption continued well into the 1980s. In 1982, for instance, the Report of the Federal Cultural Policy Review Committee (Applebaum-Hébert Report) noted that "cultural policy" was a relatively new phrase in the Canadian lexicon and pointed out that if cultural policies were merely means to other ends, economic and political, for instance, then this would cheapen them (Canada 1982, 8). At the same time, the commission's distinction between industrial and non-industrial cultural practices opened the door to considerations of industrial cultural policy (Dowler 1996, 344). Paul Audley's *Canada's Cultural Industries* (1983)—a groundbreaking study both in Canada and around the world—focuses on industrial dependency but falls into a trap of equating Canadian content with something inherently superior. Emile McAnany (1986) points out that "although Audley makes reference to the cultural or Canadian content of the industry he discusses, he never makes clear what he means by 'culture' or 'Canadian content.' Nor does he make any suggestions about how Canada might encourage cultural production other than by providing the economic incentives for doing so" (19). Audley, a private consultant, was commissioned to write the study for the Canadian Institute for Economic Policy; it is a landmark of Canadian cultural economics.

By the 1980s, dependency theory had migrated from being a leftist critique to becoming mainstream in Canada (see Smythe 1981). Paradoxically, corporate speech was just as fond of this perspective as were critical political economists. Far from rupturing the nation-building project, corporate cultural expression, such as advertising, has consistently been one of the strongest creators of national identity, which makes sense when one considers that corporations have had the most to gain economically from policies of cultural protectionism (Beaty and Sullivan 2006; Cunningham 2003, 16; Wagman 2002). Having traded the language of revolution for reformist democratic discourse using the category of "cultural citizenship," the left had effectively offered up a resonant language for appropriation by corporations and the state.

Culture, like everything else in capitalism, is clearly tied to economics. But it is an industry with a difference: the link to both the consciousness and desires of the population is also at stake. In the media reform school of communication, what is at stake is ideology, or the power to define the world. Another school of thought that has made inroads into the study of cultural policy and institutions, if not industries, is that of "governmentality," derived from the work of philosopher Michel Foucault (Bennett 1998; McGuigan 2003; Sterne 2002).

No matter what type of politics it is tied to, the impulse to engineer society through culture has roots in the logic labelled by Foucault as "governmental," and Canada is no exception. Governmentality, as Foucault uses the term, refers to the conduct of conduct, the various modes utilized to manage the subjectivity of citizens in state projects (Foucault 1991). Policy, as Tony Bennett (1998) points out, is etymologically and historically linked to policing with its intention to regulate and manage the population. Justin Lewis and Toby Miller (2003) put it this way in their introduction to the *Critical Cultural Policy Studies Reader*: "cultural policy is...a site for the production of cultural citizens, with the cultural industries providing not only a ream of representations about oneself and others, but a series of rationales for particular types of conduct" (1).

While neoliberal logic pushes for the marketization of all cultural and public goods, a governmental logic implicates the state in managing citizens biologically and culturally. And, indeed, the two can be combined, as they are in the present constellation. As Mitchell Dean (1999) explains, currently there is a "reconfiguration of the social as a set of quasi-markets in services and expertise...inflected with themes of community and identity" (6). Governmental logic extends to consumer behaviours, bringing affective dimensions of the constitution of the self in line with the imperatives of the market.

Jonathan Sterne (2002) has argued that the discourse of governmentality sidelines an analysis of corporate power in favour of an emphasis on the state, a heuristic that makes it useful in small states with a fair amount of state regulation of the cultural realm (such as Canada) but ultimately limited in providing a critique of the power of capital (see also Dowler 1996). However, Lewis and Miller's (2003) suggestion that "the state often works to defend certain corporate interests, even while officially eschewing its potential to create cultural possibilities that go beyond corporate requirements" (4) certainly rings true in the Canadian case. As we shall see, the neoliberal and the governmental are only partially able to provide explanatory frameworks in a context where multiple objectives are being articulated around national culture.

In Canada, the development of anti-discrimination language paralleled the push to deregulate the media marketplace. Starting in the mid-1980s, with a mandate granted by Brian Mulroney's Progressive Conservative election victory in 1984, the Department of Communication (DOC) undertook a full review of its mandate and embraced the principle of the free flow of information. The result was new parliamentary acts in broadcasting, copyright, and telecommunications, all of which worked to converge telecommunication infrastructure and cultural industries with a set of shifts to market-driven strategies for the cultural sector, opening up the possibilities for the free trade agreements that followed (Comor 1991, 249). At the same time that the market was being endorsed as a viable cultural force, principles of diversity were being enshrined in the Canadian Constitution (1984) and in the *Multiculturalism Act* (1988). Interestingly, the Free Trade Agreement signed with the United States in 1988 compels both parties to coordinate their copyright and retransmission laws, effectively undercutting the DOC's stated commitment to the availability of a wide selection of Canadian products. Thus, while institutionalizing a number of "cultural dependencies," the DOC continued to deploy the language of cultural self-determination and "nation-building" (Comor 1991, 258).

Yet, arguably, Canada's underperforming cultural industries were the result not of too much protectionism or the cherished dependency perspective, but rather of a longstanding commitment to the free flow of information through the construction of communications infrastructure compatible with that of the US. As one commentator noted in the 1980s, "No country in the world probably is more completely committed to the practice of free flow in its culture and no country is more completely its victim" (Anthony Smith cited in Audley 1983). So perhaps the shift in language found in DOC policy in the 1980s and 1990s was not so much a shift in policy as a rationalization and intensification of a process that had begun much earlier. The difference—and it is a significant one—was in the framing. Rather than hold publicly funded cultural institutions accountable for fending off American cultural industries, the market was itself being proposed as the solution to the problem. This neoliberal direction offered to collapse cultural policy into the development of the cultural industries themselves. This marked a shift from an earlier view of Canadian cultural institutions and industries as serving the common good as well as particular national policy goals.

A new field of cultural economics emerged in this period with a group of Canadian scholars committed to endorsing the potentials of free trade and debunking unquantifiable rationales for cultural funding. Economists such as Stuart McFadyen and Colin Hoskins served as free trade supporters in position papers written for UNESCO ("Import/Export: International Flow of Television Fiction in North America," 1990) and SSHRC ("Cultural Development in an Open Economy," 1994), and produced many other publications (for a listing see McFadyen, Hoskins, and Finn 2000). These economists from the University of Alberta argued (Hoskins, Finn, and McFadyen, 1997) that audio-visual industries could and should be assessed in terms of their "cultural externalities," quantifiable measures of value (Sinclair, 1996). Keith Acheson and Christopher Maule, economists at Carleton University, took up the torch in their work arguing for the value of the free trade of cultural commodities. They posited that Canada should give up trying to create national culture and focus instead on international trade. According to them, given its comparative advantage in the English-language film marketplace, there was no point in trying to compete with Hollywood (2005b). They believed that the marketplace had spoken and that rather than helping to form cultural taste, it merely responded to it (2001, 2005a).

Acheson and Maule's work is in line with that of cultural consultant and urban development guru Richard Florida who in his 2002 publication *The Rise of the Creative Class* posed the possibility that cities might cultivate their development and tax strategies to become international hubs of creativity. Like the cultural economists discussed above, Florida's ideas about creativity and culture are entirely reducible to economic indicators and quantitative measures, and are concomitantly divorced from other ways of imagining non-market value for creativity and culture (see Blum 2010). John Hartley argues that individual creativity and industrial scale have become merged in the new media technologies that underpin the knowledge economy. Consumer, culture, market, and citizenship are all converging in what he terms the "creative industries," a policy direction taken by Australia and Singapore, among other jurisdictions (Hartley 2005, 5).

Conclusions

I began this discussion by asking researchers of the cultural industries in Canada to think about the very constitution of their object. As we have seen, the cultural industries have traditionally been considered to be forces of American mass commercial culture against which the Canadian state used cultural policy to create cultural institutions with identity-producing influence (Dowler 1996). Profit-oriented Canadian cultural industries existed, but they were not considered to be an important part of the creation of distinctive Canadian culture. From the late 1960s through to the 1980s, a critique of the inherently elitist and Eurocentric aspect of this perspective combined with a new capitalist discourse of populist diversity through the marketplace to constitute a neoliberal governmental discourse of free trade and cultural diversity. This dovetailed into the intensification of globalization, the inception of the digital economy, and the resultant centring of cultural expression for the economy. Responsibility for culture was devolved in two directions: upwards to international agencies such as the World Trade Organization, and downwards to provinces and municipalities able to encourage businesses and individuals through tax incentives and marketing campaigns (Hesmondhalgh 2007, 2).

While the cultural industries were initially seen to be the expression of capital, they grew to become paradigmatic of a new economic model—flexible, creative, precarious, meaningful, the heart of neoliberalism's affective economy. No longer used as an attempt to scale up the reach of legitimate culture through technology, the new cultural industries enliven our intensively networked era. They still demonstrate what Adorno identified as the close connection between culture and administration, though today individuals often create their own content through both intentional and casual means (Andrejevic 2002). As Miège (1989) puts it, "the cultural and communication industries are not solely a new field for the valorization of capital, they also very directly participate in social reproduction, in ways which are only now taking shape and therefore are barely perceptible. Adorno saw the death of art in these industries; he didn't suspect their participation in the reproduction—and restructuring—of Western capitalist societies" (13).

While cultural policy continues to be an instructive site to think through governmental logic, today's cultural industries have collapsed content and commerce into the everyday expressions of identity and community. Earlier analyses of the content of media representations are less germane than are the modes by which Canadians' everyday lives have been integrated into international digital networks underwritten by some of the largest companies in the world—Apple, Google, and Amazon, for example. National cultural industries have been subsumed into international flows of digital data while the neoliberal language of individual choice and responsibility through marketization has done an end-run around the desire for reform based on national policy frameworks.

There is politics in naming. Any research into the cultural industries in Canada, then, must determine which speakers are using which terms for what ends. What are the goals of the media industries and how are they articulated to any other (non-media) industries involved in large-scale meaning-making underwritten by corporate power? How is the bald-faced "intention to sell" that Horkheimer and Adorno identified operating in the particular configuration of technology and culture that characterizes the digital economy? Are the tensions between different meanings of the word "culture" apparent in the industries being studied and the policies that continue to regulate their operation? Who is benefiting from the insistence on a national framework and how is the nation itself being represented? What kind of diversity exists in the media industries themselves and in their products? On balance, we need to consider the logic of corporate power that continues to structure the cultural industries and the governmental logic of control and improvement, both of which characterize the neoliberalizing cultural scene.

References

Acheson, Keith, and Christopher Maule. 2005a. "Canada—Audio-Visual Policies: Impact on Trade." In *Cultural Diversity and International Economic Integration: The Global Governance of the Audio-Visual Sector*, ed. Paolo Guerrieri, P. Leilio Iapadre, and Georg Koopmann, 156–191. Cheltenham, UK: Edward Elgar.

———. 2005b. "Understanding Hollywood's Organisation and Continuing Success." In *An Economic History of Film*, ed. John Sedgwick and Michael Pokorny, 312–346. London: Routledge.

———. 2001. "Canadian Magazine Policy: International Conflict and Domestic Stagnation." In *Services in the International Economy*, ed. Robert M. Stern, 395–413. Ann Arbor: University of Michigan Press.

———. 1996. "Copyright, Contract, the Cultural Industries, and NAFTA." In *Mass Media and Free Trade: NAFTA and the Cultural Industries*, ed. Emile G. McAnany and Kenton T. Wilkinson, 351–379. Austin: University of Texas Press.

Acland, Charles. 1997. "Popular Film in Canada: Revisiting the Absent Audience." In *A Passion for Identity*, ed. David Taras and Beverly Rasporich, 281–296. Toronto: ITP Nelson.

Adorno, Theodor. 1991. *The Culture Industry: Selected Essays on Mass Culture*, ed. J.M. Bernstein. London: Routledge.

Andrejevic, Marc. 2002. "The Work of Being Watched: Interactive Media and the Exploitation of Self-Disclosure." *Critical Studies in Media Communication* 19 (2): 230–248.

Arnold, Matthew. [1882] 2006. *Culture and Anarchy*. Oxford: Oxford University Press.

Audley, Paul. 1983. *Canada's Cultural Industries: Broadcasting, Publishing, Records and Film*. Toronto: James Lorimer & Company Ltd., Publishers.

Beaty, Bart, and Rebecca Sullivan. 2006. *Canadian Television Today*. Calgary: University of Calgary Press.

Bennett, Tony. 1998. *Culture: A Reformer's Science*. London: Sage.

Blum, Alan. 2010. "The Imaginary of Self-Satisfaction: Reflections on the Platitude of the 'Creative City.'" In *Circulation and the City: Essays on Urban Culture*, ed. Alexandra Boutros and Will Straw, 64–95. Montreal: McGill-Queen's University Press.

Burchell, Graham, Colin Gordon, and Peter Miller, eds. 1991. *The Foucault Effect: Studies in Governmentality*. Chicago: University of Chicago Press.

Canada. 1951, *Report of the Royal Commission on Development in the Arts, Letters, and Sciences*. Ottawa: Queen's Printer.

Canada. 1982. *Report of the Federal Cultural Policy Review Committee*. Ottawa: Queen's Printer.

Canadian Educational, Sponsored & Industrial Film Archive. Accessed June 11, 2012. http://www.screenculture.org/cesif.

Canclini, Néstor Garcia. 2005. "Multicultural Policies and Integration via the Market." In *Creative Industries*, ed. John Hartley, 93–104. Oxford: Blackwell.

Comor, E. 1991. "The Department of Communications under the Free Trade Regime." *Canadian Journal of Communication* 16: 239–261.

Conference Board of Canada. 2008. *Valuing Culture: Measuring and Understanding Canada's Creative Economy*. Accessed June 11, 2012. http://www.conferenceboard.ca/e-library/abstract.aspx?did=2671.

Cunningham, Stuart. 2003. "Cultural Studies from the Viewpoint of Cultural Policy." In *Critical Policy Studies: A Reader*, ed. Justin Lewis and Toby Miller, 13–22. Oxford: Blackwell.

Dean, Mitchell. 1999. *Governmentality: Power and Rule in Modern Society*. London: Sage.

Deuze, Mark. 2007. *Media Work*. Cambridge: Polity.

Dorland, Michael. 1996. "Introduction." In *The Cultural Industries in Canada*, ed. Michael Dorland, ix–xiii. Toronto: James Lorimer & Company Ltd., Publishers.

Dowler, Kevin. 1996. "The Cultural Industries Policy Apparatus." In *The Cultural Industries in Canada*, ed. Michael Dorland, 328–346. Toronto: James Lorimer & Company Ltd., Publishers.

Druick, Zoë. 2006. "International Cultural Relations as a Factor in Postwar Canadian Cultural Policy: The Relevance of UNESCO for the Massey Commission." *Canadian Journal of Communication* 31 (1): 177–195.

———. 2007. *Projecting Canada: Government Policy and Documentary Film at the National Film Board of Canada*. Montreal: McGill-Queen's University Press.

Edwardson, Ryan. 2008. *Canadian Content: Culture and the Quest for Nationhood*. Toronto: University of Toronto Press.

Florida, Richard. 2002. *The Rise of the Creative Class*. New York: Basic Books.

Foucault, Michel. 1991. "Governmentality." In *The Foucault Effect: Studies in Governmentality*, ed. G. Burchell, C. Gordon, and P. Miller, 87–104. Chicago: University of Chicago Press.

Gattinger, Monica, and Diane Saint-Pierre. 2010. "The 'Neoliberal Turn' in Provincial Cultural Policy and Administration in Québec and Ontario: The Emergence of 'Quasi-Neoliberal' Approaches." *Canadian Journal of Communication* 35 (2): 279–302.

Gill, Rosalind. 2007. *Gender and the Media*. Cambridge: Polity Press.

Girard, Augustin. [1972] 1983. *Cultural Development: Experience and Policies*. 2nd edition. Paris: UNESCO.

Hartley, John, ed. 2005. *Creative Industries*. Oxford: Blackwell.

Harvey, David. 2005. *A Brief History of Neoliberalism*. Oxford: Oxford University Press.

Hesmondhalgh, David. 2007. *The Cultural Industries*. 2nd edition. Los Angeles: Sage.

Horkheimer, Max, and Theodor Adorno. [1944] 1990. *The Dialectic of Enlightenment*, trans. John Cumming. New York: Continuum.

Hoskins, Colin, Adam Finn, and Stuart McFadyen. 1996. "Television and Film in a Freer International Trade Environment: US Dominance and Canadian Responses." In *Mass Media and Free Trade: NAFTA and the Cultural Industries*, ed. Emile G. McAnany and Kenton T. Wilkinson, 63–91. Austin: University of Texas Press.

Huyssen, Andreas. 1986. "Mass Culture as Woman: Modernism's Other." In *After the Great Divide: Modernism, Mass Culture, Postmodernism*, 44–64. Bloomington: Indiana University Press.

Jeannotte, M. Sharon. 2010. "Going with the Flow: Neoliberalism and Cultural Policy in Manitoba and Saskatchewan." *Canadian Journal of Communication* 35 (2): 303–324.

Lewis, Justin, and Toby Miller. 2003. "Introduction." In *Critical Policy Studies: A Reader*, ed. Justin Lewis and Toby Miller, 1–9. Oxford: Blackwell.

Marontate, Jan, and Catherine Murray. 2010. "Neoliberalism in Provincial Cultural Policy Narratives: Perspectives from Two Coasts." *Canadian Journal of Communication* 35 (2): 325–343.

McAnany, Emile. 1986. "Cultural Industries in International Perspective: Convergence or Conflict." In *Progress in Communication Sciences*, vol. 7, ed. Brenda Dervin and Melvin Voigt, 1–29. Norwood, NJ: Ablex.

McFadyen, Stuart, Colin Hoskins, and Adam Finn. 2000. "Cultural Industries from an Economic/ Business Research Perspective." *Canadian Journal of Communication* 25 (1). Accessed June 11, 2012. http://www.cjc-online.ca/index.php/journal/article/view/1146.

McGuigan, Jim. 2003. "Cultural Policy Studies." In *Critical Policy Studies: A Reader*, ed. Justin Lewis and Toby Miller, 23–42. Oxford: Blackwell.

Miège, Bernard. 1989. *The Capitalization of Cultural Production*. London: Journeyman Press.

———. 2011. "Theorizing the Cultural Industries: Persistent Specificities and Reconsiderations." In *The Handbook of Political Economy of Communication*, ed. Janet Wasko, Graham Murdock, and Helena Sousa, 83–108. Oxford: Wiley-Blackwell.

Sinclair, John. 1996. "Culture and Trade: Some Theoretical and Practical Considerations." In *Mass Media and Free Trade: NAFTA and the Cultural Industries*, ed. Emile G. McAnany and Kenton T. Wilkinson, 30–60. Austin: University of Texas Press.

Smythe, Dallas. 1981. *Dependency Road: Communications, Capitalism, Consciousness and Canada*. Norwood, NJ: Ablex Publishing Corp.

Sterne, Jonathan. 2002. "Cultural Policy Studies and the Problem of Political Representation." *The Communication Review* 5: 59–89.

Tippett, Maria. 1990. *Making Culture: English-Canadian Institutions and the Arts before the Massey Commission*. Toronto: University of Toronto Press.

UNESCO. 1982. *Cultural Industries: A Challenge for the Future of Culture*. Paris: UNESCO.

Wagman, Ira. 2010. "The Policy Reflex in Canadian Communication Studies." *Canadian Journal of Communication* 35 (4): 619–630.

———. 2002. "Wheat, Barley, Hops, Citizenship: Molson's 'I am Canadian' Campaign and the Defence of Canadian National Identity through Advertising." *Velvet Light Trap* 50: 77–89.

Wright, Robert. 2001. *Hip and Trivial: Youth Culture, Book Publishing and the Greying of Canadian Nationalism*. Toronto: Canadian Scholars' Press.

8

Critical Media Research Methods: Media Ownership and Concentration

Dwayne Winseck

In this chapter I introduce some of the tools used to study media as industries and address some the difficulties of doing good research in this area. I do so by drawing on my own experience researching ownership patterns in the telecom, media, and Internet industries in Canada. My basic premise is that we must take the media industries as serious objects of analysis. Before anything else, though, we must specify our "object of analysis" and be clear about how we will do what we intend to do.

Like everybody else, media researchers have limited time, resources, and knowledge. As a result, they must set a hierarchy of research priorities; for me, focusing on the structure, ownership, dynamics, evolution, and forces that shape the media industries are at the very top of the list (Garnham 1990). I also believe that this kind of research is essential because we live at a critical juncture in time when decisions made in the near future will constitute the media landscape for decades to come—perhaps for as long as a century, if lessons from the "industrial media" set down in the nineteenth and twentieth centuries (and just now being undone) are any guide.

To these ends, I focus on what I call the network media industries, a composite of a dozen of the largest media sectors: wired and wireless telecoms services; broadcast television; multi-channel cable and satellite pay TV; cable, satellite, and IPTV distributors; newspapers; magazines; music; radio; books; movies; Internet access; search engines; social media sites; and online news sources. While the need to establish a common frame of reference seems like an obvious thing to do, this basic step is often ignored. Without a common frame of reference, however, researchers ostensibly studying the same thing reach wildly different conclusions.

147

Beyond putting a sturdy methodological foundation into place, I will illustrate the value of this approach by addressing the following question: Have the network media industries as a whole become more or less concentrated over time? To do this, I assemble data for each ownership group in these sectors and chart the trends in market concentration over more than a quarter of a century (1984–2010) using two analytical tools: concentration ratios (CR) and the Herfindahl–Hirschman Index (HHI) (see below). Doing this also allows us to map which media segments have grown, stagnated, or declined, and to weigh in on questions about the "future of the media" as well as widespread claims that some media—particularly newspapers, television, and music—are in "crisis."

Media Concentration: A Methodological Conundrum?

In *Media Ownership and Concentration in America*, Eli Noam (2009) laments the absence of a systematic body of evidence that would allow us to create a coherent portrait of the state of media concentration in the United States. He points to two key culprits behind this reality: one, gathering such information is not easy; and two, the issue is highly politicized. Much the same can be said with respect to Canada. Indeed, Philip Savage makes exactly this point when he argues that "the media ownership debate in Canada occurs in a vacuum, lacking evidence to ground arguments or potential policy creation either way" (Savage and Gasher 2008, 295). Mike Gasher concurs, asking rhetorically, "Who is really trying to measure media concentration and its impact in an empirical way? I honestly...cannot think of anyone who does that" (*ibid.*, 295).

Why is this the case? The reasons are relatively simple. First, we can say that *concerns with media concentration are episodic.* Scholars and members of the public tend to respond to such issues as a result of policy concerns being raised in a given time and place. Second, scholarly views differ on whether concentration is a problem or not, depending on the political or ideological views of the analyst. In addition, the lack of common methods leads some to state that media are forever becoming more concentrated (Bagdikian 2004) while Goldstein (2007) and Compaine (2005), in contrast, cast their net so widely to include electronics, ICTs (information communications technologies), media, and Internet together, with the consequence that even massive media goliaths appear to be but tiny specks in a vast information universe.

Of course, there is much room for disagreement as to what to include and exclude from analysis, but it is essential to clearly delineate both the scope of the terrain and the time period to be analyzed from the outset. During my work with the International Media Concentration Research Project (IMCR), an initiative involving more than forty researchers from around the world studying long-term trends in media concentration over the period from 1984 until 2010, we fixed on a range of a dozen or so key sectors. This list included wired and wireless telecoms; broadcast television; pay and subscription television; cable, satellite, and IPTV distribution networks; newspapers; magazines; radio; Internet access; search engines; social networking sites; and online news sources. Most of them conform to the standardized North American Industrial Classification System (NAICS), which is useful not just because it gives a common frame of reference, but also because Statistics Canada publishes annual revenue data for each industry on the basis of NAICS codes. This is an important early step in the research process because it allows us to establish the revenues for each sector before tallying up the market share of each player within them. Since the emphasis is on assessing trends over time, the data gathered cover a twenty-six year period (from 1984 to 2010), with data collected at four-year intervals.

The next step is to identify and gather data for each ownership group in each sector with more than a 1 per cent share of the market. In this regard, past inquiries and studies are excellent for identifying key players, as are industry association reports, historical accounts, newspaper columns, and so on. I use revenue rather than audience share or circulation as the basis unit of analysis because it is a consistent measure that can be used across all media sectors and is not as vulnerable to periodic changes in how, for instance, audiences are measured.

The Canadian Radio-television and Telecommunications Commission's (CRTC) *Communications Monitoring Report* has become a useful source of data over the last decade, but it is limited by several factors. Every year, the report provides data on revenues and market shares (but typically not both) for the biggest four or five players in radio and television broadcasting, cable and satellite distribution, and telecoms. However, these data ignore many players while even the data for the top players in each sector are presented inconsistently from one year to the next, making it all but impossible to create a coherent portrait over time.

Worse, the CRTC randomly discards data that are more than eight years old, crippling historical research. Finally, most of the information published by the CRTC, as well as Statistics Canada, is presented at the aggregate level to avoid disclosing "trade secrets"—a concept that is generously defined. Moreover, as I saw during the vertical integration hearings in the summer of 2011, the CRTC is under intense pressure from telecoms and broadcasting players to further reduce data disclosure levels, especially for new video-streaming, on-demand, and IPTV services.

I have met CRTC staff many times and have filed a dozen requests under the *Access to Information Act* to obtain more detailed data, but to no avail. The irony in all this is that the CRTC as well as the Department of Canadian Heritage and Industry Canada actually have vast amounts of data that could be useful to researchers, but they refuse to make these data available. Susan Crawford (2011), a professor at the Cardozo Law School and an expert on media anti-trust issues, captures the CRTC's minimalist view of its role when it comes to public disclosure when she refers to its website as being "truly primitive."

These limits mean that researchers have to go well beyond official sources in order to cobble together their own data. Corporate documents such as annual reports and financial statements, consultants' reports, and the business press as well as online sources such as Alexa.com, Comscore, Experian Hitwise, Internet World Stats, and Nielsen—some of which offer access to timely data for free—are essential sources. A more comprehensive list of sources is included as an appendix at the end of this chapter. Consultants' reports, however, can be hugely expensive, inconsistent, and tied too closely to clients' needs. For example, the *Global Media and Entertainment Outlook* by PricewaterhouseCoopers that I often use costs USD\$1,500 for a single user or USD\$6,000 for a multi-user library license. In such cases, cultivating contacts at these agencies is useful in the hope that they will provide some kind of free access. It is also a good idea to do this with people at the CRTC, Statistics Canada, and Department of Canadian Heritage, if you can.

Annual reports are an obligatory source of information for *publicly traded* companies such as Bell, Rogers, Shaw, QMI, and Astral because they are required by law to disclose certain kinds of audited information to investors. Annual reports are available on company websites and, for earlier years, from the companies themselves, university libraries, and Library and Archives Canada. Poring through such documents for hundreds of companies across long periods of time generates rich insights into and a feel for the dynamics and context surrounding specific media companies and the media industries as a whole.

Yet there are limits to these sources as well. For one, disclosure levels in Canada are not as stringent as in the United States or United Kingdom. Second, companies present data in wildly different ways that typically veer far from the standard NAICS classifications. This means that researchers

have to prise apart large and fungible categories that may make sense
for companies, but which (for example) group newspapers, books, and
magazines all under the "publishing" label, as QMI's annual reports
do. At other times, partial bits of information are given that require the
unknown bits to be filled in from other sources. This is especially true
for Internet subscribers, for example, where Bell, Rogers, Shaw, QMI, and
Telus provide information about the number of subscribers they have or
about overall revenues, but not both. Faced with the situation where there
are subscriber data but no discrete revenue data, it is necessary to use a
"rule of thumb," which in this case means taking the piece of information
that is available—number of subscribers—and multiplying it by another
number widely used in the industry: "average revenue per user" (ARPU).
Multiplying the former by the latter gives a good estimate of total Internet
revenues, which can then be subtracted from the larger category that cable
and telecom companies use to bury these figures so as to avoid giving
away "trade secrets." The number won't be perfect, but it will be close,
and if there is an error, you can take solace in the knowledge that the error
applies equally to all, without bias or prejudice.

Matters are tough when it comes to publicly traded and regulated companies, but tougher yet when they are neither, to the point that it becomes
impossible to get beyond general revenue figures for a specific sector and
determine market shares for players within it. I discovered this problem
in relation to the book, film, and music industries in Canada and have
not been able to complete the analysis of these sectors because of it. Many
media companies are *not* publicly traded, and thus whatever information
is available is scarce and not easily verified. This is a problem not just for
small entities, but also for important players such as Eastlink, one of the
dominant cable and Internet companies in the Maritimes and northern
Ontario, and the *Globe and Mail*, a privately held company owned and
tightly controlled by the Thomson family. It is another irony that while
the media hold themselves up as guardians of transparency and public
disclosure when it comes to government and the economy in general, they
are neither when it comes to their own affairs.

The further one digs, the more things emerge that require judgment for
which no ironclad rules apply. When the CRTC and the advocacy group
the Friends of Canadian Broadcasting analyze television and radio, for
instance, they base the CBC's market share on advertising revenues. They
exclude annual funding from Parliament on the grounds that the CBC
competes with the commercial broadcasters only for advertising revenue
and audiences' attention. In contrast, I use "total revenues" for every actor

in each media sector and therefore include the CBC's annual funding from the government. This means that I arrive at a substantially higher figure for its market share (18 per cent vs. 10 per cent, as the CRTC claims) and add a billion or so dollars to the television and radio industries, a total that is far larger but better suited to capturing all resources within the broadcasting "system" than the figures used by the CRTC and Friends of Canadian Broadcasting.

As you might imagine, keeping track of twenty-six years of data for a dozen or so media sectors and all of the players in each is not easy. You do not have to be a genius to do this, but learning to use a spreadsheet program such as Excel and keeping detailed notes of what you do each step of the way is essential because there is no way to keep track of all this in your head.

After the data have been collected, concentration levels are analyzed on a sector-by-sector basis and then combined into three higher-level categories: (1) the network infrastructure industries; (2) the content industries; (3) online media. I then scaffold upwards from there to give a portrait of the network media industries as a whole. At each step, I use concentration ratios (CR) and the Herfindahl–Hirschman Index (HHI) to depict levels of concentration over time.

The CR method adds the shares of each firm and makes judgments based on widely accepted standards, with four firms (CR4) having more than 50 per cent of the market share and eight firms (CR8) more than 75 per cent being considered to indicate high levels of concentration. The HHI method squares and sums the market share of each firm to arrive at a total. If there are 100 firms in a market with a 1 per cent market share each, then markets are highly competitive, while a monopoly exists when one firm has a 100 per cent market share. The following thresholds are commonly used as guides:

HHI < 1,000   Unconcentrated
HHI > 1,000 but < 1,800 Moderately Concentrated
HHI > 1,800   Highly Concentrated

Methods in Action: The Growth and Structural Transformation of the Network Media Industries in Canada

Many surprising things emerge when you pay careful attention to assembling and analyzing the evidence over time. Three things stand out for me: (1) Canada does not have a small media economy relative to global standards; (2) the media economy grew immensely between 1984 and 2010; (3) media concentration levels have generally increased since 1984, with some important exceptions.

The idea that Canada does not have a small media economy rings strange to Canadian ears, but it has the eighth largest in the world, right after Italy and ahead of South Korea and Spain (PWC 2003, 2009, 2010; IDATE 2009). This casts doubt on the widely assumed article of faith that Canada's small media market requires big media companies to compete on a global scale.

Figure 8.1 shows that the media economy has grown immensely over the last quarter century, from $19.7 billion in 1984 to $56.1 billion in 2000, and to $68.7 billion in 2010 ("real dollars"). Even when you bracket aside wired and wireless telecom services (because they tend to overshadow everything else on account of their size) and focus just on the Internet-centric and traditional media elements in the mix, a similar pattern emerges. Indeed, these ten sectors grew greatly from $12.1 billion in revenues in 1984 to $23 billion in 2000, to $33.8 billion in 2010 (in real dollars).

Figure 8.1 The Growth of the Network Media Economy, 1984–2010

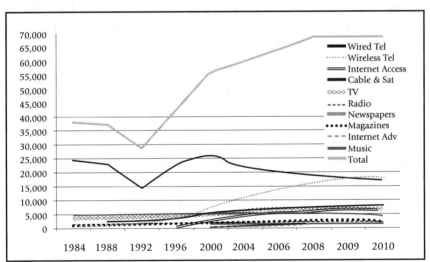

Source: Author's compilation based on sources identified in references and appendix, and methodology described in the chapter.

The emergence of pay television and wireless cellphone services since the 1980s, followed by Internet access and Internet advertising thereafter, has more than tripled the size of the media economy over the past quarter of a century. Wireless cellphone service alone is now an $18 billion sector, whereas it did not exist in 1984. Wired line telecom revenues (excluding Internet access) have fallen sharply by a third from $26 billion in 2000 to just under $17 billion last year, although this has been more than offset by gains from wireless and Internet access. Even with those sharp declines, revenues for the "connectivity industries" (wired line, wireless, Internet access) have risen sharply from $25.9 billion in 1984 to $35.3 billion in 2000, to just under $42 billion last year. Indeed, the ascent of the Internet has contributed mightily to growth in both the size and complexity of the media universe. In fact, Internet access rose from $239 million in revenue in 1996 to $6.8 billion last year. Internet advertising started from next-to-nothing in 2000 to become worth $2.2 billion last year, as well.

Trends depicted in figure 8.1 for the content industries also collide with commonplace claims that core traditional media—television, music, newspapers, and books—are teetering on the brink of calamity. For instance, broadcast television revenues—the poster child of a media sector in distress—did drift downwards from a high of $3.6 billion four years ago to $3.4 billion last year, but this is hardly catastrophic. When we take a bigger view of the television landscape, we can see that pay and specialty cable channel revenues doubled in the past decade to $3.5 billion. Combine this with the even more significant growth in television distribution—cable, satellite, and other "online video distributors" (OVDs)—and the total television universe doubled in size between 1984 and 2000, then grew again to nearly $15 billion last year. Television remains at the heart of the digital media universe.

Growth for the media economy, however, has stagnated since the economic downturn caused by the global financial crisis (which began in 2007). This is typical of a long-term historical tendency for the fate of the media economy to hinge on the state of the economy in general, however; this can also be seen in figure 8.1 where, in contrast to the comparatively milder recession years of the early 1990s, total revenues for all media in 1992 were down 20 per cent from four years earlier.

The impact of the current economic downturn varies. Revenues have largely stagnated since 2008, but have risen substantially in some cases (wireless, cable, satellite, OVDs, Internet access, Internet advertising) and fallen slightly in others (radio, magazines, music). When it comes to newspapers and wired telecom services revenues, however, there is no doubt: revenues have dropped sharply. The news is not all bleak, however, in light of the rise in newspaper advertising revenue (3.7 per cent) as well as readership in 2010, with younger readers seeming to be picking up news online via social media sites such as Facebook (Canadian Media Research Consortium 2011). Most of the "pure" newspaper publishers— Torstar, Transcontinental, *Globe and Mail*, and so on—are also recovering and some new commercial and non-commercial publications have also emerged, including, for instance, *The Mark*, *The Tyee*, *Huffington Post*, *rabble.ca*, *Straight Goods News*, *The Blogging Tories*, and *The Dominion*. In other words, while things may be tough for the industry, they might not be so bad for democracy and the range of opinion available.

The Two (or Three) Waves of Media Consolidation, 1984–2010

While the media economy has grown in both size and structural diversity, it has become, for the most part, more concentrated. In the 1980s and early 1990s, consolidation took place mostly among players in single sectors. Conrad Black's takeover of the Southam newspaper chain in 1996 epitomized the times, as did the late-1990s amalgamation amongst local ownership groups that created the large national companies that came to own the main commercial television networks—CTV, Global, TVA, CHUM, and TQS. While weighty in their own right, these changes did *not* have a significant impact *across* the media as a whole. Significant diversity still existed within sectors and across the TMI (telecom-media-Internet) sectors. The CBC remained prominent, but its share of all resources in the television "system" slid from 46 per cent in 1984 to half that amount by 2000 and about 18 per cent today.

While gradual change defined the 1980s and early 1990s, things shifted abruptly by the mid-1990s; since that time, two (and maybe three) waves of consolidation have swept across the network media industries:

Wave 1—1994 to 2000: Rogers acquires Maclean-Hunter (1994), with a peak in mergers and acquisitions from 1998 to 2001 as BCE acquires CTV and the *Globe and Mail* ($2.3 billion); Quebecor takes over Videotron, TVA, and the Sun newspaper chain ($7.4 billion) (1997–2000); and Canwest buys Global TV ($800 million) and the Hollinger group of newspapers, including the *National Post* ($3.2 billion).

Wave 2—2006 to 2007: Bell Globe Media rebrands as CTVglobemedia, while BCE exits the media business. CTVglobemedia acquires CHUM assets (MuchMusic, Citytv, and A-Channel), but with the CRTC requiring it to sell Citytv stations—which are acquired by Rogers (2007). Astral

Media buys Standard Broadcasting. Quebecor acquires Osprey Media, a mid-size newspaper chain (2006). Canwest, with Goldman Sachs, buys Alliance Atlantis (2007) (the owner of a fleet of cable and satellite channels, including Showcase, National Geographic, HGTV, BBC Canada, etc., and also the biggest film distributor in Canada).

Wave 3—2010 to present: Canwest's bankruptcy leads to its newspapers being acquired by the Postmedia Network and most of its television assets being acquired by Shaw. BCE makes a comeback and buys CTV.

Mergers and acquisitions in the telecom, broadcasting, and Internet sectors began in earnest in 1994, but mounted steadily to unheard-of levels by 2000. This trend came to a brusque halt when the TMT bubble collapsed in 2000, but rose significantly again between 2003 and 2007, before falling off sharply after the onset of the global financial crisis (ongoing since 2007). These patterns in Canada closely track those in the US and globally (Winseck 2011). It is still too early to tell whether Shaw's and Bell's (and Postmedia's) acquisitions in the past year-and-a-half constitute a third wave or whether they are just tidying up the wreckage left by the bankruptcy of Canwest.

The idea that consolidation occurs in waves shows that there is a certain periodicity to the run of events rather than a relentless process of either greater competition (as optimists like to believe) or more consolidation (as some critics claim). The telecom-media-Internet industries are not immune to consolidation and, in fact, they were destinations for capital investment at a rate far out of proportion to their weight in the "real economy" during the dot-com years. The results yielded a new entity at the very heart of the network media economy: the media conglomerate. Altogether, four massive media conglomerates, together with a half-dozen large but more specialized companies half their size, constitute the "big 10" media firms in Canada, as table 8.1 shows.

Seen from the vantage point of the "big 10," the media have become more concentrated than ever. Their share of all revenues (excluding telecom services) between 2000 and 2010 hovered steadily around 71–75 per cent—a substantial rise from 63 per cent in 1996, and an increase further still from 56 per cent in 1992. The levels are roughly two-and-a-half times as high as those in the US, based on Noam's analysis in *Media Ownership and Concentration in America* (2009). In other words, the telecom-media-Internet ecology has grown much larger and more structurally differentiated, but the "big 10" players' share of it has become significantly larger.

Breaking the picture down into the following three categories and applying the CR and HHI tools provides an even better view of long-term trends: "network infrastructure" industries (wired and wireless telecom services, Internet service providers [ISPs], cable, satellite, and other online video distributors); the "content" industries (newspapers, TV, magazines, and radio); and "online media" (search, social, and operating systems). The following paragraphs outline the evidence for each of these and then conclude with an analysis for the network media industries as a whole.

Table 8.1 The "Big 10" Media Companies in Canada, 2010 ($ millions) (excluding wired line and wireless telecoms)

	Ownership	Market Share of All Media (%)	Total $	Cable & Sat. Dist.	Internet Access	Total TV	Radio	Press/ Mags
Bell/CTV	Diversified	16	5,175.2	1,676	1,407.80	1,801.3	290.1	
Shaw	Shaw Family	15.2	4,957.2	2,331.8	916.4	1,469	240	
Rogers	Rogers Family	11.7	3,825.7	1,830	842	802	213.1	138.6
QMI	Péladeau	9.3	3,027.9	982	644.3	345.2		1,056.4
CBC	Public	4.9	1,593.4			1,235.1	358.3	
Postmedia	Godfrey et. al.	4.1	1,350.2					1,350.2
Cogeco	Audet (60%), Rogers (40%)	2.9	938.5	594.9	281.7	61.9		
Astral	Greenberg	2.7	888.1			550	338.1	
TELUS	Diversified	2.1	679.4	60	619.4			
Torstar	Atkinson, Thall, Hindmarsh, Campbell, Honderich	1.5	490.2					490.2
Total		71	32,360	8,100	6,800	6,847.6	1,910	6,502

Source: Author's compilation based on sources identified in references and appendix, and methodology described in the chapter.

The Network Infrastructure Industries

All sectors of the network media industries are highly concentrated and always have been, although Internet access is a partial exception, as shown in table 8.2.

Table 8.2 CR and HHI Scores for the Network Infrastructure Industries, 1984–2010

CR	1984	1988	1992	1996	2000	2004	2008	2010
Wireless Telecom				100	97.3	100	99.8	96.4
Wired Telecom	96.7	95.6	96.2	89.3	83.8	86.3	81.9	79.5
Cable & Sat. Dist.	35.3	54.1	69.1	78.9	80.9	85.6	85	84.2
Internet Access				33.6	54.2	53.8	54.6	56.1
HHI								
Wireless Telecom		5,591.7	5,050	5,238.7	2,675.6	3,379.9	3,246.6	3,041.3
Wired Telecom	5,034	4,431	4,193.1	3,333	2,617	1,883	2,725	2,855
Cable & Sat. Dist.	848.4	1,189.3	1,780.6	2,288.2	2,008.7	2,130	2,051.4	1,984.1
Internet Access				591.9	974.5	1,239.4	926	967.7

Source: Author's compilation based on sources identified in references and appendix, and methodology described in the chapter.

As table 8.2 shows, the CR4 (65.3) and HHI (1,883) measures for wired telecoms fell during the late 1990s as greater competition took hold. CR levels reached their lowest point ever in 2000 before the TMT bubble burst and wiped out many of the new rivals with it (CRTC 2002, 21). Competition became feebler until decade's end as a result.

The wireless telecoms have also consistently been highly concentrated, and still are, despite the advent of three newcomers last year: Mobilicity, Wind Mobile, and Public Mobile. Two competitors—Clearnet and Microcell—emerged in the late 1990s and managed to garner 12 per cent of the market between them, but were acquired by Telus and Rogers in 2000 and 2004, respectively. It is still too early to tell whether the new-comers of last year will fare any better, but with only 0.6 per cent of the market as of 2010 they are a long way from the high tide of competition set a decade ago.

As the telecom and Internet boom gathered steam in the late 1990s, new players emerged to become significant competitors in Internet access, with four new companies taking more than a third of the ISP market in 1996: AOL (12.1 per cent), Istar (7.2 per cent), Hook-Up (7.2 per cent), and Internet Direct (6.2 per cent). The early competitive ISP era, however, yielded to more concentration in the next decade. Although the "big 4" ISPs accounted for a third of all revenues in 1996, by 2000 the number had grown to 54 per cent, and it has climbed slightly higher since. As table 8.2 shows, HHI scores for ISPs are still low relative to most other sectors, but this is more of an indicator of the limits of the HHI method in this particular case, since 94 per cent of high-speed Internet subscribers rely on one or another of the incumbent cable or telecom companies' ISPs to access the Internet (CRTC 2011a, 138).

Canada has developed a framework where it is mostly incumbent tele-com and cable companies that compete with one another, not just for Internet access but for video distribution, too. In fact, cable and satellite distribution is one of the only sectors where concentration has risen stead-ily and steeply from low levels in the 1980s (850) to the top of the scales in 1996 (2,300), before drifting slightly downwards by the turn of the century to the low 2,000s, where it has remained ever since.

The Content Industries

Until the mid-1990s, all aspects of the television industry were moder-ately concentrated by HHI standards and significantly so by CR measures. Competition and diversity made some modest inroads until the late 1990s and early 2000s (depending on the measure), when the trend abruptly reversed course and has been climbing steadily ever since.

The big four television companies—Bell (CTV), Shaw (Global), Rogers (Citytv), Quebecor (TVA)—now control about 78 per cent of *all* television revenues, up from 71 per cent two years ago and the low 60s for the rest of the decade. Shaw's takeover of Canwest's television assets and Bell's buy-back of CTV pushed the levels to new extremes by 2010. Include the two next-biggest entities—the CBC and Astral—and the biggest six groups account for over 90 per cent of the *entire* television industry. The HHI scores reinforce this view.

As with the cable industries, there has never been a moment when press

Table 8.3 CR and HHI Scores for the Content Industries, 1984–2010

CR	1984	1988	1992	1996	2000	2004	2008	2010
Pay & Spec. TV		61.6	61.8	62.8	51	67.6	71.3	84.5
Conv. TV	67.2	66.7	65.8	67.9	79.9	81	80.6	81.1
All TV	64.1	61.3	57.5	56.9	64.1	61.2	71.1	77.5
Radio	49.6	44.8	50	47.9	52.2	55.5	58.8	56.4
Press	66.1	68.3	70.3	74.3	73.9	73.1	77.3	77.1
Mags	38.8	46	40.9	30.4	32.1	24.8	20	20
HHI								
Pay & Spec. TV		1,140	1,306.7	1,390.4	857.6	1,385.1	1,588.3	1,945.1
Conv. TV	2,554.1	2,066.9	2,001.1	1,819.4	1,840.8	1,939.2	1,929	2,080.7
All TV	2,307.5	1,799.8	1,522.1	1,328.8	1,243.7	1,207.1	1,519.4	1,705.8
Radio	1,257.3	935.7	1,171.9	1,042.6	924.3	961.1	1,006.6	922.6
Press	1,451.3	1,487.3	1,536.9	2,183	1,791.1	1,643.7	1,819.3	1,861.8
Mags	490	684	563	335	383	217	160	160

Source: Author's compilation based on sources identified in references and appendix, and methodology described in the chapter.

diversity has flourished. Consolidation rose steadily from 1984, when the top four players accounted for two-thirds of all revenues, to 1996, when they accounted for nearly three-quarters—a level that has stayed fairly steady since, despite periodic shuffling amongst players at the top. Magazines, in contrast, are the least concentrated, with concentration levels falling by one-half on the basis of CR scores and two-thirds for the HHI over time. Radio is also one of the most diverse media sectors according to HHI scores, but slightly concentrated by the CR4 measure.

Online Media

So far, there's little reason to believe that core elements of the Internet are immune to high levels of concentration, as measures of the ISP segment showed. But what about other core elements of the Internet: search engines, social media sites, online news sites, and browsers?

The trends are clear. Google dominates the search engine market, and this dominance is *growing*. By 2010, Google accounted for 82.4 per cent of searches. Microsoft (5.9 per cent), Yahoo! (4.2 per cent), and Facebook (1.5 per cent) trail far behind, yielding a CR4 of 94 per cent and an HHI of 6,844.5. Social media sites display similar characteristics, with Facebook accounting for 63.2 per cent of time spent on such sites in 2010, trailed by Google's YouTube (20.4 per cent), Microsoft (1.2 per cent), Twitter (0.7 per cent), and News Corp.'s Myspace (0.6 per cent) (Experien Hitwise Canada 2010). The CR4 score of 86 per cent and HHI score of 4,426 thus reveal high levels of concentration. Similar patterns emerge for web browsers, with Microsoft Internet Explorer (52.8 per cent), Google Chrome (17.7 per cent), Mozilla Firefox (17.1 per cent), and Apple Safari (3 per cent) combined having over 90 per cent market share (Comscore 2011).

Similar patterns exist with respect to the top ten websites in Canada, with the amount of time spent on such sites nearly doubling from 20 to 38 per cent between 2003 and 2008, and with most of the top fifteen online news sites belonging to well-established media outlets: CBC.ca, QMI, CTV, *Globe and Mail*, Radio Canada, *Toronto Star*, Postmedia, and Power Corp. CNN, BBC, Reuters, MSN, Google, and Yahoo! account for most of the rest (Comscore 2009; Zamaria and Fletcher 2008, 176). It is interesting to note trends in the US, however, where the most recent data collected as part of the IMCR project show that online news source concentration has gone down significantly since 2007. The data for Canada show, in contrast, a very modest decline over the same period (Fletcher 2012).

The Network Media Industries as a Whole (excluding wired and wireless telecoms)

Combining all the elements together (albeit without telecom services, for reasons explained earlier) yields a bird's-eye view of long-term trends for the network media as a whole. As figure 8.2 shows, the HHI score across all of the network media industries is not high by the criteria set out earlier, but the long-term upward trend is clear and significant.

While the HHI for the network media fell slightly during the 1980s and early 1990s, by 1996 trends reversed and levels were right back to where they had been a dozen years earlier. Thereafter, the number rose steadily

Figure 8.2 HHI Scores for the Network Media Industries, 1984–2010

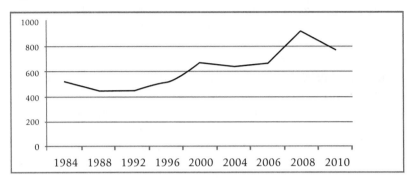

Source: Author's compilation based on sources identified in references and appendix, and methodology described in the chapter.

Figure 8.3 CR4 Score for the Network Media Industries, 1984–2010

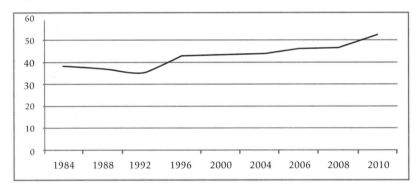

Source: Author's compilation based on sources identified in references and appendix, and methodology described in the chapter.

to 667 in 2000, where it hovered for most of the decade until rising significantly again to 780 in 2010. The CR4 standard, as shown in figure 8.3, reveals the trend even more starkly, with the big four media conglomerates—Bell, Shaw, Rogers, and QMI—accounting for more than half of all revenues in 2010, a significant rise in a vastly larger media universe from the 40 per cent that the big four had held twenty-six years earlier. Although still only moderately concentrated by the CR4 standard, this is for *all* media combined and the trend is significantly up, not down.

Concluding Thoughts

The mechanics and machinations of media industry research have many peculiar quirks and charms. It can be maddening trying to acquire the data needed to produce this kind of analysis, but incredibly satisfying to acquire such an intimate knowledge of the media as industries. The number of times that the findings which arise cut across the grain of common wisdom is a testament to the power of this research method, with three highlights standing out: Canada's media market is not small; claims about media being in crisis are overwrought; and media have generally become more concentrated, not less. All this underscores the need for theoretically informed, historical, and empirically driven research to counter the ideologically laced views and paucity of data that too often carries the day in public and policy discourses about the media in this country.

The trajectory of events in Canada is similar to patterns in the United States. Concentration levels declined in the 1980s, then rose sharply in the late 1990s until peaking around 2000, where they stayed relatively stable until rising significantly again between 2007 and 2010. The trends toward concentration have not been held at bay by the rise of the Internet and digitization, and current media concentration levels in Canada are roughly two-and-a-half times those in the US and high by global standards (Noam 2009). Moreover, large media conglomerates straddle the terrain in Canada in a manner far greater than in other advanced capitalist democracies such as the US, Germany, Japan, China, and the UK, where media conglomerates are no longer all the rage as they once were a decade ago.

The assets from the bankrupt Canwest have been shuffled in recent years and some significant new entities have emerged (Channel Zero, Postmedia, Remstar, Teksavvy, Netflix, *The Mark, The Tyee, rabble.ca, Huffington Post*). The overall consequence is that we have a set of bigger and structurally more complicated and diverse media industries, but these industries have generally become far more concentrated. There is a great deal more that can and will be said about what all this means, but in my eyes it means that concentration is no less relevant in the "digital media age" of the twenty-first century than it was during the industrial media era of centuries past.

References

Bagdikian, B. 2004. *The New Media Monopoly,* 6th edition. Boston: Beacon Press.
Canadian Media Research Consortium. 2011. "Social Networks Transforming How Canadians Get the News" (April 27). Accessed August 30, 2011. http://www.cmrcccrm.ca/en/projects/documents/CRMCSocialnewsApril27.pdf.
Canadian Radio-television and Telecommunications Commission (CRTC). 2002. *Status of*

Competition in Canadian Telecommunications Markets. Ottawa: CRTC. http://www.crtc.gc.ca/eng/ publications/reports/PolicyMonitoring/2002/gic2002.pdf>

———. 2011a. *Communications Monitoring Report.* Ottawa: CRTC. http://www.crtc.gc.ca/eng/ publications/reports/policymonitoring/2011/cmr.pdf.

———. 2011b. *Individual Pay Television, Pay-Per View, Video-on-Demand and Specialty Services – Statistical and Financial Summaries.* Ottawa: CRTC. http://www.crtc.gc.ca/eng/publications/ reports/BrAnalysis/psp2010/individual/ipsp2010.pdf.

Compaine, B. 2005. *The Media Monopoly Myth.* New York: New Millenium Research Council.

Comscore. 2010. *Canada Digital Year in Review: Canada.* Accessed August 30, 2011. http://www. comscore.com/Press_Events/Presentations_Whitepapers/2011/2010_Canada_Digital_Year_in_ Review.

Crawford, S. 2011. "Say You Were Canadian." Accessed August 30, 2011. http://scrawford.net/ blog/?s=say+you+were+canadian

Experien Hitwise Canada. 2010. "Main Data Centre: Top 20 Sites & Engines." Accessed August 30, 2011. http://www.hitwise.com/ca/datacenter/main/dashboard-10557.html.

Fletcher, F. 2012. "News Websites Most Often Visited." (Calculated from the Canadian Internet Project Data). Copy on file with the author.

Garnham, N. 1990. *Capitalism and Communication.* London: Sage.

Goldstein, K. 2007. *Measuring Media: Ownership and Diversity.* Revised Report Prepared for Canwest Mediaworks Inc. Submitted to the CRTC's Diversity of Voices Hearings.

Netmarketshare (2012). Desktop Browser Market Share. Alisa Veijo, CA: Accessed June 15, 2012. http://marketshare.hitslink.com/browser-market-share.aspx?qprid=0&qpcustomd=0&qptimefra me=Q&qpsp=49&qpnp=2#.

Noam, E. 2009. *Media Ownership and Concentration in America.* New York: Oxford University.

IDATE. 2009. *DigiWorld Yearbook 2009.* Montpellier, France: IDATE.

PricewaterhouseCoopers (PwC). 2010. *Global Entertainment and Media Outlook, 2010–14* (plus previous editions between 2010 and 2009). New York: PWC.

Savage, P., and Gasher, M. 2008. "Gaps in Canadian Media Research." *Canadian Journal of Communication* 33 (2): 291–301

Winseck, D (2011). Political Economies of the Media and the Transformation of the Global Media Industries: An Introductory Essay. In *Political Economies of the Media: the Transformation of the Global Media Industries.* Ed. D. Winseck, D. and D.Y. Jin, 3-48. London: Bloomsbury.

Zamaria, Charles, and Fletcher, Fred. 2008. *Canada Online! Internet, Media and Emerging Technologies: Uses, Attitudes, Trends and International Comparisons, Year Two Report, 2007.* Toronto, ON: Canadian Internet Project. Accessed August 30, 2011. http://www.ciponline.ca/en/ docs/2008/CIP07_CANADA_ONLINE-REPORT-FINAL%20.pdf.

Appendix: Useful Canadian Media Industries Research Resources

Annual reports by company.

Bloomberg. 2010. Bloomberg Professional. New York: Bloomberg.

Canada, Senate Committee on Transportation and Communication. 2003. *Interim Report on Canadian News Media.* Ottawa: Parliament of Canada.

Canadian Newspaper Association (CNA). *Newspaper Facts.* Toronto, ON: CNA 2000–2008 (becomes Newspaper Canada).

Canadian Newspaper (2003). *Ownership of Canadian Newspapers.* Toronto, ON: CNA.

CRTC (2011). *Communications Monitoring Report.* 2008-2011.

CRTC *Status of Competition in Canadian Telecommunications Markets* 2000-2007.

CRTC *Broadcasting Policy Monitoring Report* 2000-2007.

CRTC *Broadcast Monitoring Report* and *Pay and Specialty Statistical and Financial Summaries.*

Canadian Wireless Telecommunications Association's *Mobile Wireless Subscribers in Canada.*

Financial Post's *Survey of Industrials.*

FPInfomart's *Historical Profiles* (for each publicly-traded company).

Friends of Canadian Broadcasting. 2004. *Change in Parliamentary Appropriation.* Accessed December 4, 2008. http://friends.ca/files/PDF/CBCgrant04.pdf.

Internet Advertising Bureau Canada. (2011). 2010 actual & 2011 estimated Canadian online advertising revenue survey. Accessed September 20, 2010. http://www.iabcanada.com/wp-content/uploads/2011/07/IABCda_2010Act2011Bdg_ONLINEAdRevRpt_FINAL_Eng.pdf.

Statistics Canada. Cansim Tables for the following NAICS-defined sectors: 51111 Newspaper Publishers; 51112 Periodical Publishers; 51221-51223 Record Production, Distribution and Music Publishers; 51511 Radio Broadcasting; 51512 Television Broadcasting; 5152 Pay and Specialty Television; 51711 Wired Telecommunications Carriers; 517112 Cable and Other Program Distribution; 5172 Wireless Carriers (except Satellite); 51913 Internet Publishing and Broadcasting, and Web Search Portals.

9

Beyond Policy Analysis: Methods for Qualitative Investigation

Jeremy Shtern

Why Policy Research on Cultural Industries Matters and Can Matter to You[1]

In Canada as elsewhere, cultural industries are regulated industries. There are various forms of government activity that impact upon the structure and performance of cultural industries. For example, in the media and communication industries, licenses are required to broadcast in Canada; public service obligations force many media organizations to present publicly useful information such as news, political affairs coverage, and community programming alongside entertainment programs; content quotas exist to ensure that Canadian stories are produced for and distributed by the Canadian media; and the ever controversial issue of copyright law is part of an effort to strike an effective balance between the rights of content creators to profit from their work and the ability of audiences to enjoy, use, and modify the work of others. Furthermore, many cultural industries, in particular those in smaller countries such as Canada, combine high production costs with small markets for their final products and thus exist only because democracies have determined that their specific value to society transcends their profitability and have pushed (or allowed, to varying degrees) governments to use regulation as a means of ensuring their ongoing existence in spite of economic realities.

Generally, therefore, we can say that the role of policy in the cultural industries is to ensure that they meet social objectives alongside economic ones. Anyone who studies the cultural industries engages regularly in some form of at least implicit policy analysis. From time to time, however,

1 Special thanks to Sylvia Blake for her diligent and insightful research assistance.

consultations are convened, legislation is reformed, and policy controversies emerge. Research cultural industries long enough and you will
eventually be presented with the opportunity to shift from implicit policy
researcher to explicit policy researcher.

Not exactly what you had in mind for your cultural industries research?
You are not alone (Abramson, Shtern, and Taylor 2008). For students and
faculty alike, policy research can often seem to be an impenetrable world
dominated by technocratic jargon. Even more damning is the impression
that policy research shifts the focus away from the artistic people, creative
products, and cool environments that permeate cultural industries while
moving toward the markedly less sexy world of policy wonks, legislation,
and glacially paced governmental reform.

Conducting policy research on cultural industries is all about power—
who has it and how it is exercised. What makes such questions particularly fascinating and important is the compound power effect involved.
Through the social impact of their products, cultural industries have
immense symbolic power and influence within societies. Debates over the
structure and control of cultural industries are, in effect, contests for who
has the (political/economic) power to wield this (cultural) power (see
Silverstone 1999). While it is true that the basement of a dingy office block
in Gatineau may not make for quite as compelling a destination as the
galas of the Toronto International Film Festival (TIFF), there would be no
TIFF without Canada's cultural policy framework. The policy world may
not be edgy or hold creative class currency in a Plateau Montréal bohem-
ian hipster sort of fashion, but in the end, what is a sexier topic than the
naked exercise of power?

In contrast to the staid technocratic image that policy analysis-based
research can invoke, qualitative research *on* the policy process puts
researchers in a front-row seat for the political theatre that shapes cultural
industries. It gives researchers a sense of being where the action is, it can
attract media attention to researchers and their work, and it creates logical
future opportunities for research projects to contribute meaningfully to the
shape and impact of cultural industries.

Furthermore, while VIP passes to TIFF and face time with George
Clooney might be in short supply, policy processes are generally open as
well as meticulously documented and archived. Access is usually relatively
easy to negotiate, even for novice researchers with relatively low profiles,
and data sets are readily available. Important stakeholders are drawn to
policy processes and are thus easy to identify, make contact with, and
corner with questions. Policy processes provide logical organization and

boundaries to research projects. More pragmatically, knowledge of policy is a potential foot in the door for students looking to leverage their research into a post-degree career in the cultural industries, while policy processes provide for myriad networking opportunities with executives from government, commercial, and artistic cultural industry organizations.

In other words, I think too many students of cultural industries ignore policy too quickly because of misguided or ill-informed preconceptions about how dynamic or how useful this kind of research can be. As should be obvious from my transparent pitch, I think it can be exciting and it can be highly valuable, in terms of both knowledge production and training opportunities. Having said that, I do see that certain students of cultural industries may give policy research its fair due diligence, decide that it appeals to them to some degree but that they wouldn't know where to start with it, and ultimately pursue some other research design. The truth is that the policies that shape regulated cultural industries as an object of research pose challenges. The research design can seem just as murky and politically fraught as the eventual object of research, and equally devoid of formal instruction manuals.[2]

I can make my pitch for a rebrand of cultural industries policy research in three paragraphs, but presenting a reflexive account of how one can (and actually does) go about conducting such research is a bigger discussion that should involve numerous voices.

Indeed, while many scholars from various disciplinary backgrounds do analysis of the policies that shape the regulated cultural industries, there are few published accounts that tell the reader how they did it. As a result, there is not much in the way of published instruction manuals for how graduate students, junior scholars looking to break in, or established researchers looking to shift their focus to the policy area might do so.

This chapter presents something of a step-by-step guide, animated with

---

2 For example, many familiar reference texts for media studies and communication research methods do not explicitly address policy analysis as a research method, the research design implications of investigating the policy process, or steps one might to take to design a policy-relevant research program (cf. Berger 1998; Lindlof and Taylor 2011; Gunter 2000; Merrigan and Huston 2009; Rice and O'Gorman 2008; Treadwell 2011; Corner 1998). Other methods texts discuss the research design implications of the end goal of policy influence without getting into the nuts and bolts of designing research on policy (cf. Deacon et al. 2007) or wall off policy as a strictly quantitative research area and exclude it from discussion of qualitative research methods for media and communication studies (cf. Jensen and Jankowski 1991). One notable exception to this tendency is the series on methodology on which Anders Hansen has collaborated. The 1998 edition offers clear definitions and contexts that are valuable to novice researchers in understanding what kind of academic research can be done based on analysis of communication policy and what steps need to be taken in designing such studies. The 2008 edition includes contributions seminal to the literature on communication policy studies, but both are articles reissued from other publication venues and as such do not take the form of a how-to guide for policy research.

the example of a hypothetical policy analysis case study, through some of the stages one confronts in designing qualitative research on communication policy processes and the decisions that need to be made in the process. I argue that cultural industry policy analysis as a research area benefits from transparency about its method and reflexivity about its own subjectivity.

For the most part, this chapter will not cover general methodology concerns common to all research, but will address the specifics of policy process research.[3] This is not intended as a best-practice reference on the right way to do this, but more a reflection of one way that it has been done, for better or for worse.

I will illustrate this discussion with an example of a policy process drawn from the recent history of Canada's media industry. The aim of using this hypothetical policy research terrain is to illustrate the range of options available to students in designing policy research on the cultural industries, to offer some pointers on the method of policy research, and to reflexively address some of the issues related to validity and subjectivity that emerge from such research as a way of instigating similar reflexivity amongst student researchers. I will start by introducing this case.

The Lincoln Committee's Study of Canada's Broadcasting Policy

The study of the policies that shape regulated cultural industries brings a researcher into contact with various institutions (regulatory agencies, consumer protection ombudsmen, government departments, private consultancies, courts, multilateral organizations, etc.) and types of policy documents (legislation, regulatory decisions, legal proceedings, constitutions, international treaties, private contracts, etc.). Cultural industry policy is made and debated in national government organizations and within multilateral organizations such as UNESCO and the World Trade Organization. It is, in effect, impossible to point to or even invent a "typical" policy proceeding to be used in animating this discussion about designing research for policy analysis. But we have to start somewhere.

In May 2001, the Standing Committee on Canadian Heritage announced that it would conduct a study of Canada's broadcasting system. The committee's 2001–2005 study of Canada's broadcast policy (or the Lincoln

---

3 Please see any of the methodology textbooks cited above for more detailed guidance on how to develop a research problematic, compose research questions and hypotheses, conduct literature reviews, and so on. It is crucial to understand that all of these steps apply and are just as important to research on specific policy processes as they are to research conducted in any other environment, even if there are perhaps a series of strategies that apply specifically to policy process research.

Committee and its report) doesn't cover all or even most of the bases, but it is a rich research terrain from which many different possible studies could have been (and could still be) designed. It is also a useful illustration of various tensions and challenges in the academic analysis of cultural industry policy that I aim to discuss in this chapter. The rest of this section offers a bit of context before we move on to animate the discussion of research methods using this example case.

The Standing Committee on Canadian Heritage is a permanent committee of the Canadian federal government's House of Commons. It is composed of Members of Parliament from the various parties holding seats in the House. The mandate of the committee is to study and report on the policies, programs, and legislation of the department of Canadian Heritage and various other agencies assigned to it. Between 2001 and 2003, the committee held a series of stakeholder consultations and public meetings across Canada and received more than 200 briefs outlining the views of various stakeholders on Canada's broadcast system. In June 2003, an 872-page final report was produced that includes a series of concrete recommendations (Canada, Standing Committee on Canadian Heritage 2003). In April 2005, the Liberal government tabled its official response to the report in Parliament. Since that time a case for the impact (or lack thereof) of the committee's work has been archived in subsequent broadcast policy decisions as well as in the reforms that have been made (and the committee-recommended reforms that have not been made) to Canada's broadcast policies.

It might not be a perfect exemplar, but it is easy to imagine that, for the period between 2001 and 2005 when the report was being developed and debated, it would have piqued the attention of students and researchers interested in a number of subjects, regardless of whether they had previously considered themselves to be interested in policy analysis. Their subjects of interests could include Canada's media industries, Canadian broadcast regulation, the political economy of Canadian media, public-service broadcasting, the effect of the Internet on traditional media, or the representation of gender and cultural diversity in Canadian media. Additionally, post-2005 and into the immediate future, analysis of the *Lincoln Report* endures as a compelling case study on the structures, politics, and power relations that define the Canadian media. Furthermore, similar exercises occur in other countries and in regard to regulated cultural industries other than broadcasting. Any number of policy analysis-based studies could be designed around this process and report. In the following sections, we will examine a handful of hypothetical ones.

Policy Document Analysis

An intellectual, normative, and methodological centre of research into the policies that shape cultural industries is the analysis of policy documents: pieces of legislation, regulatory documents, negotiated treaties and agreements, court cases, government reports, and others.

One of the advantages of policy research is the extensive document trail that policy processes produce and archive. In the case of the *Lincoln Report*, there is, for example, the extensive (to say the least) final report which includes a series of concrete recommendations. Using this report, I could structure a significant research project and report based on a research design that did nothing more than frame analysis of this text according to some predetermined theoretical and normative framework.[4]

The difficult question is what framework to use. How does one "read" policy documents in such a way that the analysis generated holds the weight of academic research? Consider the range of possibilities.

Such a framework for analysis could be established and articulated through a thorough synthesis of relevant contributions in various academic literatures. For example, a normative analytical framework for the *Lincoln Report* could be constructed that drew on insights from literatures on innovation in media industries, technological convergence, public-service broadcasting, and the political economy of communication. A framework based on this sort of exercise might interrogate how effectively the *Lincoln Report*'s recommendations managed to strike a balance between the need to incubate innovative practices within Canadian communication firms through more light-touch regulation and the interests of the average Canadian citizen in managing scarce media resources without clear public interest functions becoming monopolized and commercialized. Ideally, a comprehensive and transparent framework of discourse and knowledge such as this would form the basis of policy document analysis that, if not objective, could at least be described as rigorous and balanced.

Paradoxically, this familiar form of policy document analysis is both a highly portable and marketable skill for post-degree career options in the cultural industries and, at the same time, the likely source of many

---

4 This is not the only sort of document analysis that could be designed around a policy document such as this. For example, a research design could propose to use the *Lincoln Report* as the normative framework for analysis of existing legislation such as the *Broadcasting Act*. Such an approach could apply the principles that the research argues to represent the view of the *Lincoln Report* to analysis of existing legislative texts, or could devise an international comparative policy analysis wherein the principles of *Lincoln Report* were used as the basis for analysis of an analogs policy document from another country (or vice versa).

students' inclinations to think that policy research can be dull and techno-cratic as a task and opaque and inaccessible as a research method. It is unfortunate that many students and researchers associate the method of policy research only with this form of policy document analysis, because policy research can involve a great deal more.

One way to complement and support (or triangulate) conclusions that arise from the analysis of policy documents is to conduct more qualitative research on the processes, controversies, and debates that produce policy for the cultural industries.

Qualitative research on the processes through which policy is developed, shaped, and ultimately ratified, for example on the work of the Lincoln Committee, can involve various social and investigative data collection techniques including prolonged observation-based field research; can put researchers at the centre of issues being debated within the media and public opinion; and can allow researchers to rub shoulders with leading political, activist, and business figures in the cultural industries. In the introduction, I tried to hook you on policy research by suggesting that its central theme should be framed as power rather than bureaucracy. The next section presents the effective second phase of this pitch: that the method of policy research, though perhaps typically seen as debating semantics, can be as immersive, applied, and grounded in practice as any other type of research into cultural industries.

Investigating the Policy Development Process: Actors, Issues, and Institutions

While the composition of final policy documents is obviously explicitly important to research agendas in regulated cultural industries, the process-es through which issues emerge and decisions are taken can be a privileged research terrain from which to investigate the evolution of creative indus-tries as well as the politics that define them. Investigating how these poli-cies are developed can provide unique insight into the social, economic, and political realities that shape the cultural industries.

Such research focuses on the actors, issues, practices, and institutional frameworks that shape decisions. The aim is to identify the key trends, structures, and players that are setting the agenda for regulation of a given creative industry at a given moment and organize analysis of policy docu-ments and policy processes around a handful of key empirical markers as a way of making the research terrain more manageable.[5] The following section represents an effort to outline some of the elements and issues involved in designing research on the processes policy development for the regulated cultural industries. We start by discussing what I mean when I talk about *actors*, *issues*, and *institutions*, illustrating the discussion with our example of the Lincoln Committee.

Actors: In most policy proceedings the diversity of views can be grouped into a handful of different perspectives on any issue by following the key actors involved, identifying these shared viewpoints and strategic alliances, and charting how they evolve and change. This is an exercise in the synthesis of politics rather than segmentation by organizational attributes. Once the key perspectives and certain active, emblematic spokespeople for each have been identified, research design can focus in depth on a handful of actors in the effort to avoid being worn thin while comprehensively covering the entire process.

Issues: Research design focuses on one or more specific issues of interest that are highly significant to the policy process under study. To return to the Lincoln Committee example, the future of the CBC and public broadcasting in Canada as well as the impact of newly emerging information and communication technologies on Canada's broadcasting system were but two of the numerous issues discussed, yet they received the most attention, cross-cut most of the other discussions, and were amongst the most controversial items. A case could be made that the *Lincoln Report*, was, in many respects, defined by these two issues. The *Lincoln Report* obviously discusses other unrelated issues—making Canadian broadcasting more accessible to hearing-challenged Canadians through enhanced closed captioning, for example—but analysis of such issues would not flush out the views of the major actors and would provide a snapshot of the geometry of knowledge, interests, and power within a cultural industry at a given moment in the same way that more cross-cutting issues do.

Institutions: All of this should be contextualized by analysis of how the specific institution that is hosting the policy process is influencing its outcomes. In the *Lincoln Report* case, for instance, it would be relevant to consider the following questions: What impact did the non-binding mandate of the committee have on the report's contents? What impacts did the local rules of procedure have on the choice of items to be discussed and

5 Such approaches have provided valuable insight within literatures on the production of media policies (cf. McQuail and Siune 1986; Raboy 1990a; Raboy and Padovani 2010) and content (Caldwell 2008; Mayer, Banks, and Caldwell 2009). Similarly, actor-centred qualitative methodologies are used in social network analysis and within various strains of constructivist sociology of technology, particularly actor network theory.

omitted? Why was the study assigned to the Standing Committee and not the CRTC or even to a Royal Commission? The answers to such questions provide crucial context to policy analysis.

Method and Time: Historic vs. Emergent Policy Research

Qualitative research on the actors, issues, and institutions central to policy processes will vary depending on how the timeline of the policy development and the research program overlap. We could describe two different streams of the actors, issues, and institutions approach to policy process research: *historic* and *emergent*.

Historic policy research takes the form of an *oral history*, where the backstory of an existing set of statutes is investigated. At its best, for example in *Missed Opportunities* (Marc Raboy's [1990b] singular history of Canada's broadcast policy), such research provides an empirical and normative framework for substantiating and expanding upon analysis of the policy itself with the investigation of certain details and perspectives on how it came to be. Data collection in this strain of actors, issues, and institutions policy research on the *Lincoln Report* would involve conducting semi-structured interviews with surviving participants; consideration of whatever audio/video recordings of proceedings, transcripts, summaries, and press coverage or published firsthand accounts of proceedings might be available; as well as mining the archives of the various government organizations conducting hearings and the stakeholder organizations that were participating.

With *emergent* policy research, the researcher goes into the field before there is a policy document to analyze and chronicles the incubation of a new policy, including all of the tensions and debates that define its eventual form. The opportunity to conduct emergent research would have presented itself to any researchers interested in communications law and policy, the Canadian television industry, or other relevant topics during the lead-up to the Lincoln Committee starting its work in 2001 or 2002. The same set of semi-structured interviews and archival and document analysis described above would still be central to the research design, but in contrast to the historic method, here participant observation of the committee's hearings themselves would be at the centre of the data collection strategy. In this case, follow-up participant observation at the parliamentary debates on the report would also be relevant.

The choice between the historic and emergent strains of the actors, issues, and stakeholders approach ultimately comes down to timing and the resources that researchers are able to access. On balance, while the historic strain does not require travel and absence from other responsibilities to attend a series of policy meetings in the way that the emergent strain does, the emergent strain provides policy novices and junior researchers an invaluable opportunity to understand the context of policy making and develop a network of contacts amongst policy stakeholders.

Before the comparison goes too far, it is worth noting that in many cases, emergent actors, issues, and stakeholders research may include some historic dimensions. In the case of the *Lincoln Report* example, it is easy to imagine a researcher being made aware of and drawn to the process by media attention or a colleague's interest in the early stages of the hearings. At that point, an emergent design could be used to investigate the subsequent phases of the hearing and the debates over the release of the report. In parallel, the researcher could conduct historic research using archival sources and interview data on the phases of the committee's activity that preceded his or her participant observation. In this case and generally speaking, before setting the balance between historic and emergent dimensions, it is important to consider what data will form the empirical base of the policy study and how it will be collected.

Data Collection Techniques

Policy processes are typically large, complex, highly subjective, and multi-faceted research terrains. There is no one vantage point from which one could ever claim to comprehensively capture the entire process. It is important to understand these limits but also that, in spite of them, the more volume and forms of data collected, the better. This approach can involve various combinations of virtually any established forms of qualitative or quantitative data collection techniques. However, in my own work on policy processes, I tend to make use of a specific set of data collection tools: participant observation; semi-structured interviews; and document and policy analysis. I will briefly discuss each.[6]

Participant Observation

Participant observation can provide researchers with a necessary sense of context within a complex and multi-faceted research terrain. Drawing on ethnographic research and methods developed in anthropology, the contribution of participant observation is established in media and communication research (cf. Hansen et al. 1998).

---

6 These are all best described as qualitative methods, but there is no reason why quantitative methods could not contribute to data collection as well.

By bringing together the relevant stakeholders and asking them to articulate their views and argue for their interests on a given issue, policy processes play host to their form of political theatre—and any seat with a view of the stage is revealing. In the *Lincoln Report* example, the views of various stakeholders would have come out through their interventions at various consultation sessions, as would the inclinations of the various members of the committee and the different political parties represented. Observation of such meetings is valuable for bringing the research into contact with the exercise of more implicit politics; relationships between various stakeholders and evidence of backroom dealing, for example.

Most governance agencies based on democratic principles are open to some form of citizen participation (or at least make some allowance for accrediting stakeholders from the public at large).[7] In certain cases, scholars can be invited as high-level experts. Elsewhere, access for an academic researcher may need to be negotiated with organizers in advance. In many cases where accreditation is required, attaining it simply involves submitting a general request or registering on a website. Many CRTC proceedings ask only that you confirm your attendance with a staffer in advance by email, and the UN Internet Governance Forum (IGF) has an online registration system open to everyone. In other cases, requests for accreditation are highly involved and formalized, requiring a vetting process.[8]

Yet participant observation of policy issues readily extends beyond the conference room walls. Though this was unlikely to have been the case during the *Lincoln Report* period, in certain cases policy stakeholders now make use of open Internet tools to share information, lobby each other, collaborate, and often gossip. Individuals involved in high-profile policy debates may grant media interviews, write articles about their experiences and viewpoints (for example, Mehta, Kaushik, and Kaukab 2012), or even maintain active personal blogs or Twitter feeds. Monitoring all of these can contribute to a more comprehensive observation of the actors, issues, and stakeholders.[9]

Working Document Analysis

The governance process produces an astounding volume of official and internal documentation that goes far beyond the final reports, treaties,

---

7 There are exceptions, of course, even in the most democratic societies. For example, industry self-regulatory bodies such as the Canadian Broadcast Standards Council (CBSC) are often not open to public participation or observation.

8 This is the case for many UN and multilateral organizations.

9 The existence of such publicly available "in their own words" statements can also provide a back door to being able to quote stakeholders who may not agree to being interviewed or who will only agree to unattributed or off-the-record interviews.

or policy statements. This corpus represents the most instructive and most abundant resource of information on the different actors and their approaches to communication policy making.

In the Lincoln Committee scenario, examples of working policy documents could include any number of the 200 or so interventions submitted to the committee by various stakeholders as well as any draft reports circulated, press releases, and other responses to the final report. While all the same challenges of rigour, subjectivity, and validity that were discussed in regard to final policy documents apply to analyzing working policy documents, the analysis of dozens (if not hundreds) of working documents alongside the handful of final outcome documents draws on a much larger empirical basis and is informed by the full range of ideas that were proposed.

Hard copies of such working documents are often circulated in the room during meetings. It is now common practice for governance organizations to eventually archive all official and working documents on websites, but historical research on policy developed prior to the Internet age may require visits to agency archives.

Semi-structured Interviews

Research interviews are a central component of most approaches to qualitative research and are accepted as a valid standalone methodology themselves (Hansen et al. 1988). The participant observation phase can be used to target a sample of key actors. Semi-structured interviews can then be conducted with this sample as a way of corroborating the preliminary conclusions that the researcher has formed through participant observation, investigating the behind-the-scenes story of the policy processes and, where participants will speak on the record, giving the researcher the option to use the participants' own words to narrate events in research outputs.

In the *Lincoln Report* example, interviews could be conducted with members of the committee, spokespeople for various major stakeholders (as well as spokespeople for important perspectives—even if they are from smaller organizations), civil servants and political staffers working on the file, expert consultants supporting the committee's work, journalists who covered the committee, gatekeepers from the industries, and organizations that would have been affected by the recommendations. In addition, it often proves valuable to conduct interviews with staffers involved in facilitating the negotiations.[10]

---

10 Interviews conducted with the staff who have been responsible for collecting, summarizing, and reporting on policy interventions provide for a comparing of notes which can function as an effective, if informal, test of inter-coder reliability.

Analysis of Data

Data collected through all of these different tracks can be assembled and analyzed over the course of the case study. This is an iterative process that necessarily involves some preliminary fieldwork before research design is complete and an effectively open attitude toward redesigning the analytical framework for the data as the research terrain evolves. It also crucially treats all of the different data sets relatively interchangeably.[11]

To use the *Lincoln Report* example, a researcher could have begun participant observation of the public consultation meetings and hearings with little more than a set of thematic interests. In that case, the observation phase would be a valuable time to revise, further develop, augment, and replace the existing themes with a much more detailed set of questions, hypotheses, and organizing themes that could form the basis of a set of interview protocols. This thematic guide could be further modified following the knowledge gained from the interviews for use as an informal coding framework for policy documents. During the analysis phase, comments expressed by certain stakeholders in policy documentation would be considered alongside any thematically similar interview excerpts and notes from participant observation.

In this sense, the framework of discourse and knowledge used to structure policy analysis is based on empirical as well as theoretical dimensions, and is contextualized to a degree that allows researchers to do at least some "reading between the lines" of the policy documents.

Beyond Policy Analysis: Reflections on This Method

In the hypothetical qualitative study of the *Lincoln Report* process, research would have followed the actors (Canadian media companies, government agencies, activists, etc.), issues (the future of public broadcasting in Canada, the impact of digital technologies), and the institutions (the standing committee that convened the proceeding; the CRTC and CBC, that would deal with its fallout; the Royal Commission that was not created, etc.). In contrast to many students' perceived view of policy research as detached and technocratic, the researcher would have conducted immersive research on the exercise of power in the Canadian cultural industries, would have become an expert on a topic that was the subject of public scrutiny and media coverage, would have developed portable policy knowledge and skills, and would have developed a network of contacts

---

11 In this sense it is associated with the literature on grounded theory (Glaser and Strauss 1967; Charmaz 2000).

amongst business, government, and activist leaders in the Canadian cultural industries. This chapter has sought to challenge some of the typical negative preconceptions that students have about policy research. I hope it is clear by now that the method of such research, when it is extended beyond analysis of documents and into the theatre of cultural industry politics, can be dynamic and rewarding. In other words, I hope that this chapter's survey of how one might conduct qualitative research on the policy process has answered the superficial, but by no means insignificant, reservations that students might have about the task of policy research. In conclusion, I would like to make a case that this discussion of qualitative research on the policy process also responds to the more substantive reservation that researchers might have about this sort of work—about its rigour and its validity.

In the policy document analysis section of this chapter, I suggested that a framework could be constructed for analysis of the *Lincoln Report* that drew on insights from a multitude of academic literatures and that such a framework could be used to interrogate how effectively the *Lincoln Report*'s recommendations managed to strike a balance between various competing interests and objectives. In the earlier discussion, I suggested that a comprehensive and transparent framework of discourse and knowledge such as this, though perhaps not objective, could at least be described as rigorous and balanced.

I want to return to this discussion in reflecting on the value of the qualitative method that was subsequently presented. This standard for policy document analysis is a slippery slope, and the definitions of balance and rigour can vary greatly depending on how broad the normative parameters are set and how reflexively they are applied. For instance, a normative framework for analysis of the *Lincoln Report* could just as easily be constructed out of nothing more systematic than a more selective reading of one set of literature. If one were to ground the normative framework for analysis of the *Lincoln Report* solely within the literature on the deregulation and liberalization of the media, analysis of the *Lincoln Report* would likely be highly critical of its invocations of citizen rights, its support for the notion of public service obligations, and, overall, its backing of continued significant government intervention in the broadcast industry. In contrast, if the normative framework for analysis of the *Lincoln Report* drew on only a handful of sources from the work of some of the more dogmatic political economy of media scholars, the policy analysis exercise might conclude that the *Lincoln Report*'s failure to pose fundamental challenges to the status quo in areas such as the concentration of media ownership, the power imbalance between community and private broadcasting, and the commercialization of public service broadcasting reflected a broad corporate and neoliberal agenda. These are two exercises of policy analysis in which a normative framework for analysis is grounded (to varying degrees) in relevant academic literature with two different sets of conclusions. Even the analyst's predispositions, own political agenda, or affiliation with a certain specific company, political party, commissioning lobby group, or activist campaign could provide the criteria used to evaluate the impacts and define the desirability of the policy document under analysis (though obviously not without seriously stretching the credibility of any claim to be offering scholarly research-based conclusions).

However established, application of the analysis framework to the policy documentation can be a similarly murky process. What matters most in policy documents are the macro implications: ideological undertones, innovative approaches, clear nods in the direction of placating certain stakeholders or lobbies over others, certain specific headline policy changes or reinforcements of a status quo that seemed to have been effectively challenged, and above all the moral force and budgetary resources put behind various statements of principle (or not). Unlike with news coverage, the airtime given to a specific issue in a policy document does not necessarily reflect its relative importance. The fact that the *Lincoln Report* includes relatively extensive discussion of the need to reinforce the community broadcasting sector directly explained neither how much of a priority the issue was for the committee nor how likely such reforms were to be made on the basis of the report. In other words, when policy document analysis is deployed as a standalone research method, strict adherence to any formal social science data analysis methodologies that use coding frameworks to count contents is usually of limited utility.

Furthermore, there is the challenge of extracting valid meaning directly from the documents alone. Legislation, negotiated agreements, and reports presented to political bodies represent codifications of compromise. Such policy documents are often written in their own lexicon wherein understanding the moral force of certain expressed principles and even the real meaning of terminology requires significant reading between the lines and immersion in the context and culture of the community of practice. At another very practical level of validity, there is the need to assure accuracy in the interpretation of documents. Most social scientists lack formal legal training and even the biggest policy wonks among us can overlook or mischaracterize precise legal minutia that matter immensely in the legalistic world of regulation.

All of which is to say that embedding the analysis of cultural industries policy documentation within larger studies of the processes through which the cultural industries are governed allows academic researchers to address many of these concerns. Collecting empirical data on the social, political, and economic context provides a rigorous framework of discourse and knowledge that can be used to frame the researcher's analysis of the policy documents themselves. By focusing on the traditional preoccupations of social science—particularly communication and cultural studies such as institutions, processes, power relations, marginalized voices, and institutionalized creativity—researchers can contribute analysis of the macro implications of policy documents and of the "between-the-lines context" that is so crucial to reading policy documents. Such research allows social scientists to contribute to the discussion from their own specialized knowledge base and, in the process, to complement rather than (potentially) dilute the sort of micro-legalistic analysis that focuses exclusively on the documents themselves.

Like all efforts to chronicle history, communication policy outcome documents are largely written by those who are most influential over the duration of the decision-making process. Focusing on the entire process allows scholars to witness compromises being made and power differentials being exploited. Such research, in other words, can often lay bare the power relations and politics that define political economies of communication. In many cases the policy alternatives that were posed, considered, and eventually rejected or fatally compromised will be more interesting than the policies that are adopted—and even more revealing about the political economies and power relations that define the role of regulated cultural industries in society. In the longer term, having had such close, prolonged personal engagement with the decision-making process better equips junior scholars to conduct policy analysis in the future by giving them a sense of the context in which specific pieces of legislation are produced. In other words, taking this qualitative, process-focused approach to policy analysis shifts social scientists from a space more appropriate for lawyers and policy wonks directly into their comfort zone of macro social tensions.

References

Abramson, B.D., J. Shtern, and G. Taylor. 2008. "'More and Better' Research? Critical Communication Studies and the Problem of Policy Relevance." *Canadian Journal of Communication* 33 (2): 303–317.

Berger, A. 1998. *Media Research Techniques*. 2nd edition. Thousand Oaks, CA: Sage.

Caldwell, J. 2008. *Production Culture : Industrial Reflexivity and Critical Practice in Film and Television*. Durham, NC : Duke University Press.

Canada. Standing Committee on Canadian Heritage. 2003. *Our Cultural Sovereignty: The Second Century of Canadian Broadcasting*. Ottawa: Communication Canada.

Charmaz, K. 2000. "Grounded Theory: Objectivist and Constructivist Methods." In *Handbook of*

Qualitative Research, 2nd edition, ed. N.K. Denkin and Y.S. Lincoln, 509–536. London: Sage.

Corner, J. 1998. *Studying Media.* Edinburgh: Edinburgh University Press.

Deacon, D., M. Pickering, P. Golding, and G. Murdock. 2007. *Researching Communications: A Practical Guide to Methods in Media and Cultural Analysis.* 2nd edition. London: Hodder Education.

Glaser, B.G., and A.L. Strauss. 1967. *The Discovery of Grounded Theory.* Chicago: Aldine.

Gunter, B. 2000. *Media Research Methods: Measuring Audiences, Reactions and Impact.* London: Sage.

Hansen, A., ed. 2008. *Mass Communication Research Methods.* Vol. 2. London: Sage.

Hansen, A., S. Cottle, R. Negrine, and C. Newbold. 1998. *Mass Communication Research Methods.* New York: New York University Press.

Jensen, K.B., and N. Jankowski, eds. (1991). *A handbook of qualitative methodologies for mass communication research.* London, UK: Routledge.

Lindlof, T., and B. Taylor. 2011. *Qualitative Communication Research Methods.* 3rd edition. Thousand Oaks, CA: Sage.

Majchrzak, A. 1984. *Methods for Policy Research.* Newbury Park, CA: Sage.

Mayer, V., M. Banks, and J. Caldwell. 2009. *Production Studies: Cultural Studies of Media Industries.* New York: Routledge.

McQuail, D., and K. Siune. 1986. *New Media Politics: Comparative Perspectives in Western Europe.* Beverly Hills: Sage.

Mehta, P., A. Kaushik, and R. Kaukab. 2012. *Reflections from the Front Line: Developing Country Negotiators in the WTO.* Delhi: Academic Foundation.

Merrigan, G., and C. Huston. 2009. *Communication Research Methods.* 2nd edition. New York: Oxford University Press.

Raboy, M. 1990a. "Le rôle des acteurs dans l'élaboration de la politique canadienne de la radiodiffusion." *Communication et information* 11 (2): 251–271.

———. 1990b. *Missed Opportunities: The Story of Canada's Broadcast Policy.* Montreal: McGill-Queen's University Press.

Raboy, M., and C. Padovani. 2010. "Mapping Global Media Policy: Concepts, Frameworks, Methods." *Communication, Culture & Critique* 3: 150–169.

Rice, J., and M. O'Gorman, eds. 2008. *New Media/New Methods: The Academic Turn from Literacy to Electracy.* West Lafayette, IN: Parlor Press.

Silverstone, R. 1999. *Why Study the Media?* London: Sage.

Treadwell, D. 2011. *Introducing Communication Research: Paths of Inquiry.* Thousand Oaks, CA: Sage.

10

Using Production Studies to Analyze *Canada: A People's History*

Olivier Côté

Television remains, among the Canadian cultural industries, strangely neglected as an object of study. This is due in large part to improper and indiscriminate use of various methodological communication approaches. These approaches rarely offer an adequate integration of television, film, and cultural studies. Some notable exceptions: Mary Jane Miller made it a priority to perform textual analyses of many works of English-Canadian television fiction series (Miller 1987, 2008). Some publications offer remarkable analysis of television-related genres (reality shows, documentaries, soaps, etc.) (Hogarth 2001, 2002, 2008; Varga 2009; Druick 2008; Byers 2005: Sloniowski 2002) and put the history of Canadian television into context (Rutherford 1990; Couture 1989; Laurence 1978, 1981, 1982; Clarke 1987).

Production studies, whose emergence as a disciplinary field is quite recent, could help eliminate restrictive disciplinary boundaries that are detrimental to the development of media studies. The field is the sum of various approaches—ethnographic, sociological, critical, materialist, political economy—and the combined effort of a variety of disciplines— anthropology, media studies, cultural geography, film studies, and sociology. The interpretative paradigm of cultural studies, which sees culture as a place of power dynamics, gives it coherence (Mayer et al. 2009).

Journalists are among the professionals whose practices have been the most studied. Critical journalism studies[1] are particularly keen to study

---

1 Several journals are the main vehicles of "critical journalism studies": *Journalism Quarterly; Journalism: Theory, Practice and Criticism;* and *Journalism Studies.* See also Zeziler (2004, 2006); Berry (2000); Carter et al. (1998); Allan (1999, 2005); Allan and Adam (1995); Allan and Zeziler (2002); Evans (2002); Deuze (2005); Meltzer (2009, 59–74).

how a common and dominant cultural knowledge, rooted in the professional conscience of journalists, helps to strengthen their social power, to legitimate political and economic forces of the established order, and, first and foremost, to maintain the cultural hegemony of the elites. This cultural knowledge, the direct result of the professionalization of journalists, has organized itself around values and concepts they broadly share. These values define and legitimize journalistic norms and practices in the storytelling of news coverage. Key journalistic principles include:

- public service, or the role of journalists as the watchdogs of democracy and the disseminators of information;
- journalistic objectivity, which consists of impartiality, neutrality, fairness, and soundness of report;
- journalistic independence (integrity);
- capacity to report major live events;
- adherence to ethical values formalized in ethical codes of conduct (Deuze 2005, 444–447; Hall 1987, 360).

Some Quebec media scholars take an organizational studies approach in their analysis of journalistic practices. These journalistic practices, they discover, are standardized by professional and social networks, organizational culture, and economic and political forces. The formation of a journalistic authority is considered in terms of strategies implemented by one or many journalists to ensure social control in the newsroom. The focal point of Quebec scholars is Radio-Canada and, in terms of TV genres and types, TV news reports, elections coverage, and live reports (Brin 2000, 2003; Charron and De Bonville 2005; Leclerc 2000).

In sum, with the exception of recent work by Serra Tinic, the integrated approach of "production studies" is still very marginal in the study of non-news TV production, both in the Anglo-Canadian and Franco-Quebecois worlds, even if its pertinence in communication and media studies has been established (Levine 2007; Mayer, Banks, and Caldwell 2009). Moreover, studies on the cultural industries in Quebec and English Canada make the presumption that Quebec's and English Canada's cultural industries operate in isolation, largely cut off from industrial developments in the other linguistic half of the country.

The TV series *Canada: A People's History* represents an interesting case study to explore the possibilities of production studies and the intersection between Quebec's cultural industries and those of English Canada. The seventeen-part docudrama was conceived of by journalist Mark Starowicz,

was produced in the two official languages in both Montreal and Toronto
(1995–2002), and was broadcast between 2000 and 2002. As such, this is
an interesting case study of four types of tensions:

a) the institutional tension between the CBC and Radio-
    Canada;
b) the tension between assumed audience expectations
    and the implied viewer;
c) the tensions caused by the power dynamics between
    executive producer Mark Starowicz, senior editorial
    team members, and production teams;
d) the tensions around the issue of "journalistic
    impartiality."

Adopting a production studies approach, including comparative analysis of original and final scripts as well as a study of nineteen interviews
with the producers, this chapter focuses on these four tensions.

Production studies is also an appropriate methodology because textual
content of television broadcasts cannot be addressed without understanding the ever-changing motivations of TV producers and their attempts to
reconcile the conflicting expectations of implied audience and commercial
goals (Tinic 2005, xi).

Production Studies in Canada

What is the place of production studies in Canada, especially in Quebec,
with respect to TV? The rise of cultural studies is quite recent in the francophone world. In France, it was not until the late 1990s with the publication in French translation of Stuart Hall's and Judith Butler's founding
texts, and with the creation of dedicated research centres, that francophone
scholars finally integrated cultural studies into media studies, both con-
ceptually and methodologically (Kaenel et al. 2003; Glevarec et al. 2008).

A partial cultural studies integration occurred in Quebec in the 1980s
among students of media studies at McGill University and in a joint
Université de Montréal-Concordia-UQÀM doctoral program. Ultimately,
however, cultural studies was never fully incorporated into media studies
in the Quebec francophone world (Yelle 2000, 2009). As a result, the "political economy" of cultural industries remains the main research paradigm,
as opposed to that of "cultural studies" (see, for example, Martin 1988).

Although the place of cultural studies is more prominent in English-
Canadian media and sociology departments (Wilfrid Laurier, McGill

University, and the University of Alberta, to name a few), most television studies programs focus on the public policy of broadcasting, including the cultural
mandate of CBC/Radio-Canada (Dowler 1996; Raboy 1996; Collins 1990;
Smith 1998; Dorland 1996; Filion 2006; Fournier 1999; Litt 1992) and, more
generally, on the political economy of the TV medium (Beaty and Sullivan
2006), failing to empirically analyze how TV programs are produced locally.

There are also few studies that offer interviews with TV program producers (including Friesen 2003; Miller 1996; Nash 1996; Stewart 1986;
Frum 1990, 1996; Koch 1986; Starowicz 2003). None of these, however,
reflect the fact that producers' discourse is a non-critical discourse, which
reinforces the cultural industry's centric representations and dramatizes
the narration of production.

Few Canadian studies propose a microhistorical analysis of TV programs and films in an ethnographic perspective, following the example
of production studies. Archivist and historian Monica Macdonald (2002)
carried out a doctoral thesis on the challenges of the CBC production of
historical documentaries from 1952 to 2002 that focuses on interpretative
negotiations between historians and journalists. This is a good example
of an empirical approach to television. Another notable example is the
study of sociologist Serra Tinic in her book *On Location: Canada's Television
Industry in a Global Market* (2005), which analyzes the conditions of producing made-for-export TV programs and films in Vancouver ("Hollywood
North"). Her extensive interviews with cultural producers are used to trace
the matrix of cultural, professional, and local practices of television production. This microanalysis of local forms of power fits more broadly into
a macroanalysis of the role hegemonic forces of the market and Canadian
cultural policy (constraints related to Canadian content, purchases of cultural products by the CBC, etc.) play in the content of cultural products
located in a regional space, but which are required also to universalize
their content based on the expectations of the global market.

Like Serra Tinic, production studies scholars study professional organizations from a wide range of professional categories. They investigate the
formation of informal networks of solidarity, communities of practice,
and their norms. Members of a production are therefore considered cultural actors reconfiguring their identities according to their social interactions
and political and economic changes.

Canada: A People's History: Background

Canada: A People's History (CAP) is the most ambitious documentary
television program ever produced by CBC/Radio-Canada, both in terms

of money invested (approximately $30 million) and the extent of the Canadian past that it covered (extending from the country's Aboriginal origins to the present).

CAP was born just after the 1995 referendum on Quebec's separation. Mark Starowicz, a CBC producer and strong supporter of the cultural mandate of public television, felt at the time that a "televised history" of Canada would be the perfect antidote to the linguistic fragmentation of the Canadian communication world. According to Starowicz, Anglo-Canadians and Franco-Canadians belong to two different communication worlds but are both subject to the commodification and the Americanization of Canadian television. Senior management of the CBC welcomed the project enthusiastically in the context of a new millennium and an interest in rediscovering Canada's past. Jim Byrd, vice president of the CBC, gave his official approval to the project on March 21, 1996 (Macdonald 2008, 240–241). Under the leadership of senior producer Mark Starowicz, the project was set in motion; production teams were formed both in Montreal and Toronto. These teams had to wait four years before they saw the first fruits of their labour.

The TV series *Canada: A People's History* was broadcast from 2000 to 2002 in two phases. The first nine episodes were devoted to pre-Confederation history (15,000 BC–AD 1873), and ran from October 22, 2000, to February 18, 2001, with a lull during the holiday season. They were mainly broadcast on Sundays from 8 PM to 10 PM. CBC and Radio-Canada departed every now and then from this choice, opting for an irregular schedule that even avid viewers found confusing. The show also benefited from reruns on Newsworld and RDI, the English and French-speaking CBC's news channels. The first phase of episodes drew over two million viewers per episode, including reruns (Allen 2001).

The second phase of episodes (Episodes 10–17) dealt with post-Confederation history (1873–1990), was broadcast in the fall of 2001 at the CBC, and ran over two seasons at Radio-Canada (September 2001 and January 2002). As a result of extensive media coverage of September 11, viewers of the CBC needed to wait until Sunday, September 30, to watch Episode 10. Broadcast of successive episodes (Episodes 11–17) lasted until November 18, one episode per week, from 7 PM to 9 PM. At Radio-Canada, Episodes 10 to 13 were broadcast on the first three Sundays of September. Viewers needed to wait until Sunday, January 6, 2002, for the full sequence of episodes (Episodes 14–17). According to Mario Cardinal, this late broadcast was due to the need to translate Toronto's episodes into French (Cardinal 2005, 171–172). For journalists, low ratings of the late

episodes resulted in low profitability of advertising, which in turn would
have delayed the broadcast of Episodes 14–17. As a result of this resched-
uling, ratings dropped to an all-time low average of 650,000 viewers per
episode (Atherton 2002).

Use of the TV series in Anglo-Canadians schools—85 per cent of them
owned it on DVD or video (Schmidt 2002)—breathed new life into *CAP*.
Additionally, beginning in 2004, Omni Television, Toronto's multicultural network, broadcast *CAP* in five languages.

Canada: A People's History offers a broad picture of Canadian history. It
created an eclectic mixture of past nation-building and new multiculturalism: a Canadian wilderness, a political country founded by Great White
Men ("the Founding Fathers"), and the passage from colony to nation
on one side; on another, a land with Indigenous and migrant forebears.
Its narrative is structured around political and military events, placing
social history in the margins and taking into account the dualistic relations and tensions between French and English people: the Seven Years'
War (or War of the Conquest) (Episode 4: "Battle for a Continent"); the
Canadian-American War of 1812 (Episode 5: "A Question of Loyalties");
the Rebellions of 1837–1838 (Episode 7: "Rebellions and Reform");
Confederation (Episode 8: "The Great Enterprise"); the battles in Western
Canada between the federal government and the Métis (Episodes 9
and 10: "From Sea to Sea" and "Taking the West"); Canadian participa-
tion in the First and Second World Wars (Episodes 11, 12, 13, and 14:
"The Great Transformation," "Ordeal by Fire," "Hard Times," and "The
Crucible"); and the 1980 referendum and the failure of the Meech Lake
Accord (Episode 17: "In an Uncertain World"). A large component of its
narrative is focused on the contribution of successive waves of migrants
and Aboriginals, which add up to the Canadian multicultural mosaic: the
First Nations of North America (Episode 1: "When the World Began"); the
permanent settlement of the French in New France (Episodes 2, 3, 4, and,
to some extent, 6: "Adventurers and Mystics," "Claiming the Wilderness,"
"Battle for a Continent," and "Pathfinders"); the influx into Canada of
Loyalists from various backgrounds, which marks the beginning of English
Canada (Episode 5: "A Question of Loyalties"); the installation, in the late
nineteenth century, of Eastern Europeans in Western Canada (Episode 11:
"The Great Transformation"); and Aboriginal people who reclaim their
rights in the 1970s (Episode 16: "Years of Hope and Anger").

Because of its scope and its impact, *Canada: A People's History* has been
extensively covered in the academic literature. First and foremost, historians and journalists involved in the production of the series wrote articles

about their participation (Starowicz 2003; Allen and Cullingham 2001; Allen 2001; Robert 2003; Cardinal 2005). Joe Friesen's efforts to interview producers of the series should be acknowledged (Friesen 2003), as should the groundbreaking effort of historian Monica Macdonald (2008) to analyze the context of its production. There are also the numerous studies of historian scholars focused on a historiographical comparison (Dick 2008; Frank 2003; Conrad 2001; Groulx 2001), as well as other studies in art history and communication, among others, which put into perspective the text of the series according to a poststructuralist approach (Hobday 2006), a teaching approach (Bryant and Clark 2006), or a psychoanalytic approach (Brooke 2002). It should be noted that none of the above studies take into account the four types of tensions previously identified.

CBC and Radio-Canada Institutional Tension

CBC/Radio-Canada's organizational structure severely undermines its overarching cultural mandate. Starting with the Massey Commission, the CBC/SRC was entrusted "to invent a national culture for a society torn apart by its regional interests and threatened by Americanization" (Nielsen 1994, 158). *The Broadcasting Act* (1991) states that CBC/Radio-Canada is "a public service essential to the maintenance and enhancement of national identity and cultural sovereignty," and that public television must contribute "to share a national consciousness and Canadian identity."[2] This policy aims to emphasize and unify Canadian identity references in the mass media and thus, ultimately, to symbolically colonize the Canadian collective imagination (Manning 2003, xix).

Most journalists interiorize the cultural mandate of public television without coercion. Many of the production members of *CAP* distanced themselves from the interpretation that the series was part of a nation-building project, proof of their unconscious introjection of the national component of CBC/Radio-Canada's cultural mandate, or perhaps of their willingness for self-justification. Production members put forward their independent and sincere will to reproduce the historical reality of the Canadian past, and also to design a program that makes a useful contribution by illuminating the current issues of society and therefore fits more directly in the public service mission.[3]

However, CBC/Radio-Canada has two headquarters, which are an

---

2 See the website: http://www.cbc.radio-canada.ca/about/mandate.shtml.
3 Interview with Gene Allen, September 10, 2007; interview with Production 1, November 30, 2007; interview with Mario Cardinal, November 29, 2007; interview with Production 11, September 15, 2008.

organizational reflection of Canadian biculturalism. The sociologist Greg Nielsen has amply demonstrated in his study on CBC/Radio-Canada (1930–1960) that a common body of symbols takes on different meanings in francophone Quebec and anglophone Canada, first and foremost because of their separate coexistence, and secondly because of their evolution in the same institutional system. Thus, argues Nielsen (1994), "albeit subject to a single monolithic federal government...their social discourses convey [the] values, [the] norms and world views specific to their group. Both societies are relatively separated. At the same time, on a symbolic level, their statements are profoundly inseparable, as they have the same institutional history" (56, 162).

The mismatch between CBC/Radio-Canada's cultural mandate and the actual circulation of social discourses can also be explained by the distinct functions of its French and English branches. Radio-Canada's main function is to "maintain the expression of the established national culture," a function rooted in a dualistic and bicultural view of Canadian reality and based on the language policy of bilingualism. Meanwhile, CBC is trying to invent a national culture for a country sharply divided by its regions and cultures, a mandate fully consistent with the objectives of the new liberal and multicultural model of cohabitation (Nielsen 1994, 59). This dichotomy compromises the unity of social and historical representations at the two networks.

This duality of representations is reflected in CAP's senior editorial team members' historical conceptions. If five members shared about the same historical narrative in the course of their acculturation in the CBC (Toronto) and their internalization of its cultural mandate, one member who mainly worked at Radio-Canada (Montreal) put forward a singularly post-colonialist Franco-Quebecois historical narrative. This situation is also peculiar to their development as individuals in distinctive linguistic communities (Anglo-Canadian, Franco-Quebecois) and their adherence or opposition to the liberal Canadian discourse.

Assumed Audience Expectations and the Implied Viewer

Tension between notions of assumed audience expectations and the implied viewer arises from this particular culture. In fact, assumed audience expectations ("audience-reaction") is a key component of the cultural mandate of the CBC (see Allen 1998). This concept is understood as a pluralist commitment by the production members to meet the expectations of a fragmented target audience—in other words, people of different ethnocultural, social, and regional background over the age of twelve who have minimal knowledge of Canadian history (Starowicz 2003, 269). This pluralist commitment is rooted in broadcasting law, which stipulates that public television is obliged to represent the cultural and regional diversity of TV audiences[4] and, furthermore, to conform to international and Canadian journalistic practices which are becoming more aware of regional, ethnocultural, and religious pluralism (Deuze 2005, 452).

In contradiction—or in addition—to this empirical viewer, the production members must comply with an implied viewer, meaning an audience imagined by themselves and implied in the text. The main consumer, viewed chiefly from the English-speaking side of production, is implied to be a patriotic Canadian viewer sharing multicultural values. This viewer believes implicitly in the equal membership of all ethnocultural groups in the Canadian state and in the enhancement of the multicultural heritage of Canadians. Production members in Montreal alternatively imagine another implied viewer, one concerned about the survival of its culture and the presentation of its peculiar or national character on TV. These distinct implied viewers are coherent with the separate cultural mandates of CBC and Radio-Canada discussed in the last section.

Inspired by assumed audience expectations, new journalistic practices, and a liberal imagining of the implied audience—which was very close to their own dominant values and ideas—members of the senior editorial team and directors wrote the series' historical narrative in a pluralistic and regional way, without coercion from their bosses. The members I interviewed spoke of this plural, level-headed, and interested integration, and were particularly concerned about alienating a subcategory of the potential audience and offending the implied viewer:

> Everyone knew that we had a lot of interests to please, without watering the wine of Canadian history...we were conscious of having to be aware of the important divides across the country and to reflect the whole country. I didn't want us to stick in a piece of Alberta's event just to have Alberta. Fortunately, the country is vivid enough. There were things happening everywhere. You didn't represent every part of the country in every episode, but on balance...[5]

---

4 See CBC/Radio-Canada website: http://www.cbc.radio-canada.ca/about/mandate.shtml
5 Interview with Kelly Crichton, September 17, 2008.

But that's where I found the biggest struggle is, to
make sure…it is broadcasting to, not just the two
audiences, language audiences, but audiences across
the whole country. The special challenge of a series like
this is to make sure that people feel they are in there
somewhere…it's really a matter of communicating to
people, to audiences. So that they find it interesting, so
that they recognize themselves. So that's the challenge.
So that's why you need two teams.[6]

I think there was a consciousness of not insulting anyone
[emphasis added]. I don't think it was written in any
particular way to influence anything, but I think it was
written in a way not to insult anybody.[7]

This multiculturalization and regionalization of the story is neither
accompanied by an intimate knowledge of journalists from different cultural and regional backgrounds, nor of the constructed nature of racial and
ethnocultural representations, which are major challenges facing the new
multicultural journalistic practices (Deuze 2005, 453).

Power Dynamics between Mark Starowicz and the Production Teams

Another set of tensions in the production develops from the unequal
power dynamics between the executive producer and members of the
production teams, and between production team members themselves. As
executive producer of the series, Mark Starowicz sat at the top of the chart.
He was the boss, and other members of the senior editorial team were
executives who owed him their current jobs and sometimes their careers,
with the exception of two journalists from Radio-Canada (Montreal). The
following comments attest to this: "He [Mark Starowicz] is like the general
of an army. We were his senior officers."[8] "Mark Starowicz was the boss.
He had the last word and he had a pretty powerful personality."[9] "You
would express any reluctance and the end result would be…no one can
face Starowicz, you cannot confront Starowicz…And he is incredibly flexible, but at the same time—he is a rather complex character—[he has] his

---

6 Interview with Production 11, September 15, 2008.
7 Interview with Production 5, May 13, 2008.
8 Interview with Gene Allen, September 10, 2007.
9 Interview with Gordon Henderson, January 9, 2004.

own quality requirements...I have not heard anyone contesting his leadership at the end...I think that the scripts were designed to please Starowicz. It was he who had the last word. He was the visionary."[10]

Language influenced the dynamics between senior editorial team members. Fluency in the language of the Other has no doubt played in the relations between members of the senior editorial team. In fact, the meeting of the senior editorial team took place in French in Montreal and in English in Toronto. Kelly Crichton had a sufficient understanding of French, but could not speak it.[11] Gordon Henderson had not mastered French at all.[12] The other members—Mark Starowicz, Gene Allen, Mario Cardinal, Hubert Gendron, and Louis Martin—were not perfectly bilingual, but had a sufficient knowledge of both French and English.

Relations between the directors—there was fifteen of them—and the executive producer of the series were also essentially hierarchical. One member said, "It was: 'Here you have it, boss. I've showed you this episode, is it okay for you? Does that suit you?' And it was Mark that had absolutely and clearly the last word."[13] Another member said, "I started to talk, and then Mark said, 'You are a junior, you don't talk. This is not for you.' I was like, 'Okay. So that's fine.'"[14] Some directors were still able to resist his editorial requests.[15]

Ultimately, Mark Starowicz had the ability to fire the directors, which undoubtedly limited their ability to oppose him. The restrictive environment that is CBC/Radio-Canada, its organizational culture, the stylistic and screenplay approach particular to this TV series (i.e., "prescripts" and visual grammar), and the inability of some directors from auteur cinema backgrounds to comply with all of these elements could explain their inability to meet the expectations of their superiors—and their ultimate dismissal.[16]

This case of maladjustment is not unique: it is typical of social control in the newsroom, and of the effects of cultural conformity to a corporate style (Breed 1955, 109–116). In fact, the romantic notion of the artist as an exclusive master of his work is at odds with collective projects and the commercial imperatives of contemporary television (Butler 2002, 338). The production of television documentaries is the result of the tension

---

10 Interview with Production 2, April 29, 2008.
11 Interview with Kelly Crichton, September 17, 2008.
12 Interview with Production 10, September 17, 2008.
13 *Ibid.*
14 Interview with Production 7, September 10, 2008.
15 Interview with Production 4, May 14, 2008.
16 Interview with Production 8, November 30, 2007; interview with Hubert Gendron, November 29, 2007.

between the personal style of the director and television and institutional constraints (Nicholls 1990, 25; Kilborn and Izod 1997, 20–21).

Dismissal of directors was not just a matter of style, but also due to interpersonal conflicts with the executive producer or senior producers. The directors were commonly charged with delaying scriptwriting and shooting of particular episodes.[17]

The filmmakers that got fired or transferred to other production teams were usually directors who did not maintain a direct relationship with the executive producer, and/or had not yet been sufficiently exposed to CBC's organizational culture. The common denominator remained their inability to meet their commitments, according to the claim of the senior editorial team members—or, according to my analysis, to comply with the CBC's corporate culture and style.

The collaborative work of producers and directors can take an adversarial or collegial form. In the scripting and production process of Episodes 1, 6, 7, 11 and 16, a spirit of collegiality prevailed between producers and their directors, according to interviews conducted with members of the production teams.[18] Sometimes, however, producers and researchers argued against the consensual vision of directors and the senior editorial team, especially on the issue of full integration of white and non-white ethnocultural minorities. They proposed a counter-discourse, which was less multicultural and more subversive, but better grounded in the respective historical realities of ethnocultural minorities. In this situation, either the producer/researcher counter-discourse was modestly integrated into the historical narrative—this is called a successful interpretative negotiation—or the disappointed journalist left the project voluntarily, was fired for insubordination, or was transferred to a different production team.[19]

In sum, the hierarchy of institutional relationships is determined, first and foremost, by the professional profile of participants (in ascending order of precedence: researchers, producers, directors, senior editorial team members, executive producer), which in turn clearly dictates the respective influence of journalists on the story.

There are several other factors to consider in the power dynamics. First, the situation of the cultural identity of journalists within the CBC/Radio-Canada dualist institutional context—the opinion of a

---

17 Interview with Mario Cardinal, November 29, 2007.
18 Interview with Production 4, May 14, 2008; interview with Production 9, January 25, 2008; interview with Production 1, November 30, 2007; interview with Production 10, September 17, 2008.
19 Interview with Production 11, September 15, 2008; interview with Production 15, December 2003. Other journalists were also dissatisfied with their experience: Production 21 and Production 22.

Franco-Quebecois or Anglo-Canadian was not seen in the same way according to his place of work (Montreal or Toronto). Additional factors affecting power dynamics were the political positions of journalists (sovereignists, Quebec nationalists, federalists, conservatives, liberals, etc.), although this was never explicitly expressed between journalists in their relations, as well as the participants' genders and the minority expression of a feminist discourse.

Journalistic Impartiality as Source of Tension

The journalists who worked on *Canada: A People's History* proclaimed their impartiality and authority over Canada's past. They challenged the authority of historian advisors on this highly contested terrain and, as such, proved that journalistic impartiality can create tension.

As the series is journalistic in its form and in its content, journalists possessed the power to make editorial choices and were not required to convince historians of the relevance of these choices. It must be noted that since the 1970s the professionalization of CBC's journalists—mainly the sharing of the same journalistic code—has authorized their authority on the past in the production of historical documentaries. Before this, at the CBC, the past was the preserve of historian experts (Macdonald 2008, 276).

Journalist production members consistently relied on their socio-professional ideology—that is, their internalized concepts of objectivity, integrity, and independence—to establish their authority on the past.

According to the principle of integrity, the journalist must put aside his own prejudices and outside interference with the laudable goal of producing "neutral" content. The application of this principle was even more crucial as the production and broadcast of the series were made against a background of financial scandal involving the federal government. In the media, Mark Starowicz[20] and Mario Cardinal defended the integrity of the journalists who worked on the series. In addition, senior producer Hubert Gendron assumed that the largely shared principle of journalistic integrity among production members was an effective barrier against regionalism: "Overriding our regional origins is the common discipline of journalism." (Frank, Fernandez, and Fleming 2000).

This notion of journalistic autonomy could result in limiting an in-house presentation of innovative journalistic points of views, both in terms of content and form. As Deuze (2005) observes, most journalistic

---

20 Mark Starowicz, quoted in Raymond Giroux, "Financement chatouilleux," *Le Soleil*, August 14, 2000, A8.

innovations appear to compromise editorial autonomy and peer accept-
ability. As a result, editorial independence becomes "ideological value"
in that it allows resistance to change (449). Multicultural integration of
ethnocultural groups would therefore likely be severely compromised in
its most subversive form—challenges to the established order, the narration of unsuccessful attempts of migrant integration—and, moreover, this
would oversimplify the complexity of the past.

This is where the principle of inclusive equity gets taken up. Inspired by
assumed audience expectations, each constituent part of society is entitled
to the presentation of its views. Journalistic impartiality is used here as
a balancing tool to expose interpretations in a symmetrical way and,
according to cultural studies scholars, for the presentation of consensual
interpretations, tools of the status quo (Raphael 1995, 63). In the media,
Hubert Gendron described the presentation of the Seven Years' War as a
typical example of this balancing act: "If you look at the Seven Years' War,
for example, I think we did a really good job of saying 'here's how the
French see it, here's how the English see it...' *it's a history without a point of
view because it's a history of different points [of view]* and it doesn't organize
it for you or analyze it for you. It doesn't feed you a conclusion" (Friesen
2003, 195, emphasis added).

The only "gaps" in journalistic authority on the past were found in
the principles of accuracy and fairness, applied by historians who were
entrusted to verify the authenticity of documents, costumes, and historical figures as well as the veracity of the series' interpretations. According
to Production Member 9, the historian is, by definition, a fact checker
who serves journalists' needs: "I can say that we were pretty close to
the historians for checking and crosschecking facts. We tried not to say
anything."[21]

Although production members did not make explicit reference to journalistic socio-professional ideology in their in-house discussions, they did
recognize a common set of journalistic values, which replaced the ideologies of the political spectrum that are invisible in the everyday interactions
at work. As Kelly Crichton puts it, "It's a kind of social rule at the CBC that
nobody discusses its politics, nobody talks about what their political beliefs
are, they don't label themselves. Maybe people have, [but] I don't know who
of my colleagues are conservatives or liberals. I have ideas but I don't know;
we never speak about it. It is a forbidden subject like religion."[22]

At Radio-Canada (Montreal), Mario Cardinal explained that the

---

21 Interview with Production 9, January 25, 2008.
22 Interview with Kelly Crichton, September 17, 2008.

ideologies of each member were fully assumed outside of working hours and during lunchtime.[23] This shows the full extent of the hegemonic nature of socio-professional ideology. According to Gene Allen, CBC's journalistic norms "[were] our *sense of being*, having worked as journalists for a long time, how journalists approach contentious issues."[24] Starowicz added, "Whatever language you speak, whatever country you come from, *journalism is one thing that has common factors built in, and those people understand each other.*"[25] Hubert Gendron said, "We *all* know what a good story is" (Frank, Fernandez, and Fleming 2000).

Socio-professional journalistic ideology contributed to the formation of a common organizational language between producers, directors, and researchers. This code reinforced the bonds of journalists' community of interest—a community which was then required to stand together regardless of the directions and the decisions. They sacrificed their individuality through their allegiances to a collective work, largely shaped by the expectations and the writing of executive producer Mark Starowicz, breeding the maintenance of a hegemonic discourse about the past. Directors of Production 6 and Production 11 had been at the CBC for over twenty years and adhered to this journalistic code without any pressure being exerted on them by the senior editorial team: "The given is that you will apply journalistic criteria to this history story."[26] "It's in your DNA."[27]

The sharing of this code, however, built internal tension. In fact, the code was compromised by the presence within the senior editorial team of a member putting forward a feminist and leftist counter-discourse.[28] The rumours about his or her political affiliations and comments were seen as a blow to the socio-professional journalistic ideology and to the consensual interpretative order, considering journalists are informally liberal or conservative: "We had…to reopen debates that had been resolved."[29] This masculine organizational culture makes no room for political subjectivity. Feminists have no choice but to silence their beliefs if they want opportunities to get ahead and get peer recognition.[30]

Journalistic impartiality can also be a source of tension with estab-

---

23 Interview with Mario Cardinal, November 29, 2007.
24 Interview with Gene Allen, September 10, 2007.
25 Mark Starowicz, quoted in Murray Whyte, "Two Solitudes, One History…," *National Post*, January 10, 2001, B03.
26 Interview with Production 6, May 13, 2008.
27 Interview with Production 11, September 15, 2008.
28 Interview with Editorial Team 3, September 17, 2008.
29 Interview with Hubert Gendron, November 29, 2007.
30 The example of women's treatment in Australian newsrooms confirms this. See North 2009.

lished political power. In establishing their authority, journalists seek both to please the established political power and to defend a critical and independent point of view according to the public interest, both of which belong more distinctly to CBC's organizational culture than they do to Radio-Canada—so much so that CBC's journalists are potentially the producers of consensual interpretations (dominant discourse) and oppositional interpretations. In view of the fact that public television is financed by the Canadian state in a non-recurring fashion, the governing party is likely to reduce or withdraw funding as it sees fit. This sword of Damocles applies a constant pressure on journalistic freedom of expression and impartiality (Taras 1995 737–738; Macdonald 2008, 4).

Conclusion

The producers ultimately appropriated the dominant historical discourses of society, reformulating these master narratives so that they fit into one coherent TV narrative, perfectly appropriate for *Canada: A People's History*. They are, as such, the privileged interpreters of Canadian culture.

Unequal power dynamics within the CBC/Radio-Canada led in due course to the liberal reinvention of the Anglo-Canadian narrative of the past, and on to the TV-style iteration of a supportive narrative of coexistence based on the new pluralism of Canadian society. The proposed story was at the confluence of colonial and postcolonial representations, of old-fashioned nation-building and new multiculturalism.

This consensual story of Canada's past aimed to increase the social cohesion of Canadian society. The story was one of status quo in its pacification of social conflicts and in its support of a new, state-related Canadian mode of cohabitation, both dual and multicultural. CBC's cultural producers therefore participated in the creation of a pan-Canadian "imagined community" (Anderson 1991), with *"raisons communes d'exister"* (common reasons to exist) (Dumont 1995) for a Canadian society increasingly divided and fragmented.

References

Allan, Stuart. 1999. *News Culture*. Buckingham: Open University Press.
———. 2005. *Journalism: Critical Issues*. Buckingham: Open University Press.
Allan, Stuart, and Barbara Adam. 2006. *Theorizing Culture: An Interdisciplinary Critique after Postmodernism*. New York: New York Press.
Allan, Stuart, and Barbie Zeziler. 2002. *Journalism after September 11*. New York: Routledge.
Allen, Gene. 1998. *Untitled document*, correspondence to Historian 7, March 19. CBC Documentary Unit Archives (Toronto); *Canada: A People's History*, Series Production Binder 7, Files 7, Historian Comments.
———. 2001. "The Professionals and the Public: Responses to *Canada: A People's History*." *Canadian Historical Review* 82 (1): 381–391.

Allen, Gene, and James Cullingham. 2001. "CHR Forum: Canadian History and Film." *Canadian Historical Review* 82 (1): 333-346.

Anderson, Benedict. 1991. *Imagined Communities: Reflections on the Origins and Spread of Nationalism.* London: Verso.

Atherton, Tony. 2002. "Canada Turned to CNN after Sept. 1; Canadian Networks' Viewership Changed Little." *Ottawa Citizen,* January 29, D6.

Beaty, Bart, and Rebecca Sullivan. 2006. *Canadian Television Today.* Calgary: University of Calgary Press.

Berry, David. 2000. *Ethics and Media Culture: Practices and Representations.* Oxford: Focal Press.

Breed, W. 1955. "Social Control in the Newsroom : A Functional Analysis." *Social Forces* 37: 109-116.

Brin, Colette. 2000. "L'influence stratégique des journalistes politiques dans un contexte de tension normative: La couverture électorale à la télévision de Radio-Canada." *Les Cahiers du journalisme* no. 7: 44-58.

———. 2003. "L'organisation médiatique et le changement des pratiques journalistiques: Adaptation, innovation et réforme." In *La science politique au Québec: le dernier des maîtres fondateurs,* ed. Jean Crête, 417-431. Quebec: Les Presses de l'Université Laval.

Brooke, Glenn. 2002. "Canada: A People's History—An Analysis of the Visual Narrative for a Colonial Nation." MA thesis, Concordia University.

Bryant, Darren, and Penney Clark. 2006. "Historical Empathy and Canada: A People's History." *Canadian Journal of Education* 29 (4): 1039-1064.

Butler, Jeremy. 2002. *Television: Critical Methods and Applications.* London: Routledge.

Byers, Michele, ed. 2005. *Growing Up Degrassi.* Toronto: Sumach Press.

Cardinal, Mario. 2005. *Il ne faut pas toujours croire les journalistes.* Montreal: Bayard Canada.

Carter, Cynthia, et al. 1998. *News, Gender and Power.* New York: Routledge.

Charron, Jean, et al. 2005. "De la théorie au terrain: modèle explicatif de l'évolution du journal télévisé au Québec." Quebec: Département d'information et de communication, Université Laval.

Clarke, Debra Maria. 1987. "Network Television News and Current Affairs in Canada: The Social Contexts of Production." PhD dissertation, Carleton University.

Collins, Richard. 1990. *Culture, Communication and National Identity: The Case of Canadian Television.* Toronto: University of Toronto Press.

Conrad, Margaret. 2001. "My Canada Includes the Atlantic Provinces." *Social History* 31 (68): 392-402.

Couture, André M. 1989. "Elements for a Social History of Television: Radio-Canada and Quebec Society, 1952-1960." MA thesis, McGill University.

Deuze, Mark. 2005. "What Is Journalism? Professional Identity and Ideology of Journalists Reconsidered." *Journalism* 6 (4): 442-464.

Dick, Lyle. 2008. "Representing National History on Television. The Case of Canada: A People's History." In *Programming Reality: Perspectives on English-Canadian Television,* ed. Zoë Druick and Aspa Kotsopoulos, 32-44. Waterloo, ON: Wilfrid Laurier University Press.

Dorland, Michael. 1996. *The Cultural Industries in Canada: Problems, Policies and Prospects.* Toronto: James Lorimer & Company Ltd., Publishers.

Dowler, Kevin. 1996. "The Cultural Industries Policy Apparatus." In *The Cultural Industries in Canada: Problems, Policies and Prospects,* ed. Michael Dorland, 328-346. Toronto, James Lorimer & Company Ltd., Publishers.

Druick, Zoë, and Aspa Kotsopoulos. 2008. *Programming Reality: Perspectives on English-Canadian Television.* Waterloo, ON: Wilfrid Laurier University Press.

Dumont, Fernand. 1995. *Raisons communes.* Montreal: Boréal.

Evans, Michael Robert. 2002. "Hegemony and Discourse : Negotiating Cultural Relationships through Media Production." *Journalism* 3 (3): 309-329.

Filion, Michel. 2006. "L'évolution des politiques publiques et des pratiques culturelles en matière de radiodiffusion canadienne. L'utopie et la réalité." *Globe: revue internationale d'études québécoises* 9 (2): 75-89.

Fournier, Jean-Pierre. 1999. *La télévision entre culture et commerce.* Sainte-Foy: Centre d'études sur les médias, Université Laval.

Frank, David. 2003. "Public History and the People's History: A View From Atlantic Canada." *Acadiensis* 32 (2): 120-133.

Frank, Steve, Sand Fernandez, and Thomas Fleming. 2000. "History with a Bang." *Time*, October 23, 18.

Friesen, Joe. 2003. "*Canada: A People's History* as 'Journalists' History.'" *History Workshop* (56): 184–203.

Frum, Linda. 1990. *The Newsmakers: Behing the Cameras with Canada's Top TV Journalists*. Toronto: Key Porter Books.

———. 1996. *Barbara Frum: A Daughter's Memoir*. Toronto: Random House of Canada.

Glevarec, Hervé, et al. 2008. *Cultural Studies: Anthologie*. Paris: Armand Collin.

Groulx, Patrice. 2001. "La meilleure histoire du monde." *Social History* 34 (68).

Hall, Stuart. 1987. "Media Power: The Double Bind." In *New Challenges for Documentary*, ed. Alan Rosenthal, 357–365. Berkeley: University of California Press.

Hobday, Joyce. 2006. "A Critical Discourse Analysis of *Canada: A People's History*." MEd thesis, University of Saskatchewan.

Hogarth, David. 2001. "Local Representation in a Global Age: A Case Study of Canadian Documentary Television as a Transnational Genre." *International Journal of Canadian Studies* 23: 133–155.

———. 2002. *Documentary Television in Canada: From National Public Service to Global Marketplace*. Montreal: McGill-Queen's University Press.

———. 2008. "Reenacting Canada: The Nation State as an Object of Desire in the Early Years of Canadian Broadcasting." In *Programming Reality: Perspectives on English-Canadian Television*, ed. Zoë Druick and Aspa Kotsopoulos, 17–31. Waterloo, ON: Wilfrid Laurier University Press.

Kaenel, C., et al. 2003. *Cultural studies. Études culturelles*. Nancy: Presses universitaires de Nancy.

Kilborn, Richard, and John Izod. 1997. *An Introduction to Television Documentary: Confronting Reality*. Manchester: Manchester University Press.

Koch, Eric. 1986. *Inside Seven Days: The Show that Shook the Nation*. Scarborough, ON: Prentice-Hall.

Laurence, Gérard. 1978. "Histoire des programmes de télévision: essai méthodologique appliqué aux cinq premières années de CBFT Montréal, sept. 1952–sept. 1957." PhD dissertation, Université Laval.

———. 1981. "La rencontre du théâtre et de la télévision au Québec (1952–1957)." *Études littéraires* 14 (2): 215–249.

———. 1982. "Le début des affaires publiques à la télévision québécoise 1952–1957." *Revue d'histoire de l'Amérique française*: 213–239.

Leclerc, Gérard. 2000. "L'information en direct à la télévision ou comment les journalistes adoptent de nouvelles normes professionnelles." *Les Cahiers du journalisme*: 34–43.

Levine, Elena. 2007. *Wallowing in Sex: The New Sexual Culture of 1970s American Television*. Durham, NC: Duke University Press.

Litt, Paul. 1992. *The Muses, the Masses, and the Massey Commission*. Toronto: University of Toronto Press.

Macdonald, Monica. 2008. "Producing the Public Past: Canadian History on CBC Television." PhD dissertation, York University.

Manning, Erin. 2003. *Ephemeral Territories: Representing Nation, Home and Identity in Canada*. Minneapolis: University of Minnesota Press.

Martin, Claude. 1988. "L'économie politique des industries culturelles et la prise en compte des auditoires." *Communication-information* 9 (3).

Mayer, Vicki, et al. 2009. *Production Studies: Cultural Studies of Media Industries*. London: Routledge.

McKay, Ian. 1998. "After Canada: On Amnesia and Apocalypse in the Contemporary Crisis." *Acadiensis*: 76–94.

———. 2000. "The Liberal Order Framework: A Prospectus for a Reconnaissance of Canadian History." *Canadian Historical Review* 81 (4): 617–645.

Meltzer, Kimberly. 2009. "Journalists' Perspectives According to News Medium." *Journalism Practice*: 59–74.

Miller, Mary Jane. 1987. *Turn Up the Contrast: CBC Television Drama since 1952*. Vancouver: University of British Columbia Press.

———. 1996. *Rewind and Search: Conversations with the Makers and Decision-Makers of CBC Television Drama*. Montreal: McGill-Queen's University Press.

———. 2008. *Outside Looking In: Viewing First Nations Peoples in Canadian Dramatic Television Series*. Montreal: McGill-Queen's University Press.

Nash, Knowlton. 1996. *Cue the Elephant! Backstage Tales at the CBC*. Toronto: McClelland and Stewart.

Nicholls, Bill. 1990. *Introduction to Documentary.* Bloomington: Indiana University Press.

Nielsen, Greg Marc. 1994. *Le Canada de Radio-Canada : sociologie critique et dialogisme culturel.* Toronto: GREF.

North, Louise. 2009. "Rejecting the 'F-Word: How 'Feminism' and 'Feminists' Are Understood in the Newsroom." *Journalism* 10 (6): 739–757.

Raboy, Marc. 1996. *Occasions ratées: histoire de la politique canadienne de radiodiffusion.* Sainte-Foy: Presses de l'Université Laval.

Raphael, Chad. 1995. *Investigated Reporting.* Urbana: University of Illinois Press, 2005.

Robert, Jean-Claude. 2003. "L'histoire et les medias." *Revue d'histoire de l'Amérique française* 57 (1): 57–68.

Rutherford, Paul. 1990. *When Television Was Young: Primetime Canada (1952–1967).* Toronto: University of Toronto Press.

Schmidt, Sarah. 2002. "Only One Thing Wrong with Canadian Soldier on Book Cover: He's Australian." *National Post*, March 14, A01.

Sloniowski, Jeannette. 2002. "Popularizing History: The Valour and the Horror." In *Slippery Pastimes: Reading the Popular in Canadian Culture*, ed. Jeannette Sloniowski and Joan Nicks, 159–174. Waterloo, ON: Wilfrid Laurier University Press.

Smith, Joel. 1998. *Media Policy, National Identity and Citizenry in Changing Democratic Societies: The Case of Canada.* Durham, NC: Duke University Press.

Starowicz, Mark. 2003. *Making History: The Remarkable Story Behind* Canada: A People's History. Toronto: McClelland and Stewart.

Stewart, Sandy. 1986. *Here's Looking at Us: A Personal History of Television in Canada.* Montreal: CBC Enterprises.

Taras, David. 1995. "The Struggle over 'The Valour and the Horror': Media Power and the Portrayal of War." *Canadian Journal of Political Science* 28 (4): 737–738.

Tinic, Serra A. 2005. *On Location: Canada's Television Industry in a Global Market.* Toronto: University of Toronto Press.

Varga, Darell, et al. 2009. *Rain/Drizzle/Fog: Film and Television in Atlantic Canada.* Calgary: University of Calgary Press.

Yelle, François. 2000. "Les études en communication médiatique au Québec et l'approche des *Cultural Studies*." *Commposite*: Université de Montréal.

———. 2009. "Cultural studies, francophonie, études en communication et espaces institutionnels." *Cahiers de recherche sociologique* (47): 67–90.

Zeziler, Barbara. 2004. *Taking Journalism Seriously: News and the Academy.* Thousand Oaks, CA: Sage.

———. 2006. "When Facts, Truth and Reality are God-terms: On Journalism's Uneasy Place in Cultural Studies." *Communication and Critical/Cultural Studies* 1 (1): 100–119.

11

Ethnic Broadcasting: A History

Mark Hayward

This chapter offers a brief overview of the institutional history of multicultural television in Canada. While this history discusses some of the significant actors in the development of these services and their regulation, it also makes a more pointed argument about the way in which "ethnic" television has taken shape and the broader significance of this development to discussions about media, identity, and the nation-state in Canada. This argument is that, when it comes to the development of greater cultural diversity in Canadian media (particularly with regard to television), it is essential that issues of scale be taken into account as much as issues of identity, ethnicity, or language. In other words, the engagement with questions of multicultural media should not be understood simply as the additive expansion of the languages and cultures that are available on Canadian televisions; it must also be understood as one of the sites where the CRTC and other agencies of the Canadian government have had to wrestle with questions that challenged the established scales at which media regulation took place. For this reason, in spite of being a "hidden history" in the development of media in Canada, its relevance extends far beyond its usual status as a footnote to the history of Canadian culture, often speaking to the most pressing issues regarding the relationship between media and the nation in direct and complex ways.

To support this claim, this chapter will examine two transitional periods in the evolution of multicultural television in Canada. The first of these periods involves the transition from policies that facilitated the emergence of "local" broadcasters during the 1960s and 1970s toward a policy that would allow for the creation of "national" services in the early

1980s. The second period of transition that will be discussed involves the transformations that accompanied the introduction of digital distribution platforms to the Canadian market and the subsequent expansion of access to a wide range of international, third-language services for Canadian television viewers during the early 2000s. Both of these transitions involved a rethinking of the scale at which established policies regulating multi-cultural television (and the television industry more generally) operated, troubling the established relationship between national identity, cultural diversity, and media. Thus, as much as they entailed new content being made available to Canadian audiences, they also involved a reconsidera-tion of the relationship between the local, national, and international. To examine these moments of transition, this chapter will focus on two television services: Telelatino and Fairchild TV. Both created in 1984 (as Latinovision and Chinavision, respectively), these two services will orient the following discussion of changing approaches to multicultural tele-vision in Canada. These two services were the product of the first period of transition, and they have been forced to adapt to the introduction of digital television service during the second.

However, before offering a brief survey of the existing literature, it is of perhaps fundamental importance to begin by addressing some questions of terminology. Already in this chapter, there have been a number of terms used to discuss these services, ranging from more general, abstract con-cepts like cultural diversity or multiculturalism to more common phrases like ethnic television. Rather than reduce this lexical complexity, this essay will instead seek to stick as closely as possible to the terms used in the contexts being discussed. Sometimes these changes in language speak to changes that occurred over time, as is the case in the transition from multiculturalism to cultural diversity. Other differences in terminology, especially language for referring to the television services analyzed in this chapter, have more to do with the particular venues where the discourse is taking place and, as a result, movement between the terms is considerably more fluid.

For example, analyzing the shift from multicultural discourse to the language of cultural diversity, Karina Horsti (2009) has noted that this was a transition that occurred in many sites, ranging from international organizations like UNESCO to the discourses of regional and national governments in the aftermath of a wave of negative opinion towards multiculturalism in the mid-to-late 1990s. With regard to different per-spectives on the place of media within a multicultural society, it should be noted that "third-language" media has long been a descriptive term used by government and regulatory agencies to refer to many of these services, defining them in opposition to services operating in either of the nation's two official languages (see CRTC 1985, for example). At the same time, however, phrases like "ethnic television" and "multicultural television" were in circulation. These other terms have typically been used in venues outside of official government statements and policy documents (including a range of different materials such as newspapers or publicity copy). Each of these terms resonates with a particular understanding of the relationship between television programming, cultural identity, and the nation-state in ways that should not be overlooked.

With regard to existing literature on the subject, there is a sizable body of research that addresses the subject of multiculturalism and cultural diversity in Canada. This work raises questions ranging from conceptual and theoretical discussions of the social, political, and cultural dynamics of multiculturalism to more detailed analyses of the way in which cultural diversity is practiced, whether by citizens in everyday life, by cultural producers, or politicians in clearly defined institutional settings (see Bissoondath 1994; Day 2000; Kymlicka 2001; Stein 2007; Taylor 1994). Within communication and media research, there is a growing body of literature that falls within the parameters of the research mentioned above. These include analyses of how race and cultural difference have been represented in film and television as well as discussions about the relationship between media culture and broader social trends (for example, see Roth 2005 for a study of the Aboriginal Peoples Television Network; Hirji and Karim 2009 on race and ethnicity). Finally, such research has not been limited to the university, and these issues have been addressed both by government agencies as well as organizations representing the industry (see Cavanaugh 2004; Karim 2009). There are, however, relatively few texts that offer a comprehensive history of the development of multicultural media, particularly across an extended period of time. There have been some interesting case studies offered looking at particular services and their development. However, when examined synthetically as part of a broader shift toward cultural diversity in media, these services are reduced to footnotes that are quickly passed over in the documentation of the mainstream development of the national television industry. Beatty and Sullivan, for example, make only passing reference to these services as part of a broader argument involving the nationalizing project of Canadian television policy (2006, 49–53). Other histories of Canadian television (or media more generally) barely mention these services beyond noting a few key dates and the languages in which these services broadcast (older studies make only mention of third-language services, e.g., Raboy 1990; Collins 1990).

By far, the most extensive discussions of multiculturalism in Canadian television are Lorna Roth's discussion of policies and practices relating to ethnic and third-language media published in 1998 and Ryan Edwardson's history of the development of ethnic broadcasting in Canada published in 2004. Both essays present detailed discussions of the policy frameworks that allowed third-language and ethnic broadcasting to develop in Canada and offer similar conclusions. Recognizing the considerable gains made since the 1960s, both authors suggest that the long-term success of the project to develop multicultural media in Canada is still uncertain. Roth (1998), for example, writes that the creation of these services "may, in the long run, have a positive outcome in provoking a re-evaluation of the way people think about cultural and racial differences in broader socio-politic-al contexts of everyday life." This chapter differs from both of these essays in its historical scope by extending beyond the year 2000, as well as the argument that it is the scale of media governance as much as the expansion of linguistic and cultural diversity that defines the historical development of these services to date. However, the aim of this chapter is not to suggest that these channels provide the as yet undiscovered key to understanding the development of Canadian broadcasting. Rather, it is to show that the debates and dynamics that have accompanied their development warrant more than the scant references afforded them up to this point. Rather than simply marking the ways in which particular communities have been able to advocate for their inclusion on the television spectrum, the interactions between regulatory agencies, politics, and media say something quite sig-nificant about the transformations of Canadian society and its institutions over the past fifty years.

From Local to National Multicultural Television

The emergence of ethnic media in Canada cannot be separated from the adoption of multiculturalism as a guiding concept in Canadian cul-tural policy in the late 1960s and early 1970s (see Canada, RCBB 1969). Officially adopted as government policy in 1971 in response to the con-clusions of the Royal Commission on Biculturalism and Bilingualism (1965–1969), the first period in the development of ethnic and third-language broadcasting took place between the early 1960s and the early 1980s. In 1962, the Board of Broadcast Governors (precursor to the CRTC) announced its policy regarding radio broadcasters in languages

other than English or French in response to a license application for a
station in Montreal that would serve the large number of immigrants living in the city.[1] After many years of debating the possible effects of ethnic
media on the integration of immigrants into Canadian society, the board
had decided that such broadcasters were welcome to use non-official languages if they "also provided listeners with programming on Canadian
history, politics, society and English and/or French language instruction"
(Edwardson 2008, 91). In these years, the majority of "ethnic" broadcasters were small, independent entrepreneurs operating radio services that
had little relationship with mainstream English or French language media
in Canada. Insofar as mainstream television in Canada was shaped by state
broadcasting, the CBC had explicitly declared that it would not be offering
services in any non-official languages (other than Aboriginal languages)
as it was "a federal agency, the statutory creation of Parliament...[which]
recognizes only two official languages" (cited in Edwardson 2004, 320).
The response was to allow radio stations to apply for permission to broadcast 15 per cent of their schedule in third languages (with that amount
being raised to 20 per cent in special circumstances) provided the community possessed a sufficient number of third-language speakers (typically
between 150,000 and 200,000) (Canada, RCBB 1969, 4:185). A survey
conducted in 1966 shows that there were fifty radio stations across the
country broadcasting to twenty-five different groups, constituting about
200 hours of programming total; most of these broadcasts were concentrated in Ontario and Quebec. Television constituted a much smaller portion of overall broadcast time during this same period (1966), when there
were only four hours of programming in third languages per week (more
than half of which were in Italian), broadcast in Ontario and Quebec only
(*ibid.*, Appendix Table A-150).

    With regard to the early days of local ethnic media in Canada, the career
of Johnny Lombardi and the growth of CHIN Radio are exemplary (for a
comprehensive history, see Amatiello 2009). Lombardi, a Canadian-born
member of Toronto's Italian community, moved into broadcasting as a
way of advertising his grocery store in the city's Little Italy. Initially leasing
a few hours on weekends for Italian programming sponsored by his commercial interests beginning in 1947, Lombardi attempted to launch his
own multilingual service in partnership with Ted Rogers. After the partnership with Rogers fell apart, Lombardi applied to have the frequency Rogers
had purchased transferred to his ownership and was granted the license

---

1 Board of Broadcast Governors, "Foreign Language Broadcasting," Public Announcement, January
22, 1962.

for CHIN Radio in 1966.[2] Lombardi quickly moved beyond broadcasting for the Italian community, proposing that the station carry a variety of non-English programming in numerous other languages, and within four years he was broadcasting in twenty-three languages. At the time, there were some concerns that the creation of third-language media services would slow the pace of integration and acculturation to the Canadian context. However, testifying to the CRTC to renew his broadcast license in 1970s, Lombardi framed CHIN's operation within Canadian democracy, saying, "We are providing a platform for those minority groups who, without us, would be absolutely voiceless" (cited in Edwardson 2004, 320). Characteristic of this period in ethnic broadcasting, Lombardi never expanded beyond the Toronto market; his forays into television, the weekly variety program Festival Italiano, for example, never transformed into an independent broadcast entity of its own, remaining a programming block leased from other broadcasters.[3]

Third-language broadcasting remained focused on radio broadcasting until the late 1970s when Multilingual Television Ltd., founded by Dan Iannuzzi (publisher of Toronto-based Italian-language daily Il Corriere Canadese), was granted a license in 1979 to launch a single station that would provide programming for a number of Toronto's ethnic communities (see CRTC 1986 for full details). The station, CFMT (drawing its call letters from the name "Canada's First Multilingual Television"), had a broadcast schedule that was to consist of 30 per cent English-language programming, 10 per cent French, and the remainder in one of two dozen other languages. The model was one shared with a series of multilingual radio stations that were created at the same time, and it has similarities with Australia's Special Broadcasting Service, which was launched during the same year. It was also similar to the low-power public access channel La Télévision Ethnique du Québec based in Montreal and operating since the mid-1970s.[4] While these stations catered to many different communities, they were all decidedly local in their orientation.

It was only in the early 1980s that the second stage in the translation of Canadian multiculturalism into broadcasting unfolded, coinciding with initiatives to exploit the expanded carrying capacity of cable distribution. As

2 Some details of the arrangement between Rogers and Lombardi can be found in Phil Stone (1976, 6).

3 Initially, Lombardi worked with Toronto-based Global TV (where he shared a studio with the cast of SCTV) and later moved the broadcast to Citytv in the 1980s (Amatiello 2009.)

4 While it is often recognized as an important contributor to the development of ethnic television in Canada, there is no scholarly history of TEQ currently available and very little that I could find about its operations.

part of the initiative to the create new pay-TV and cable services, a number of licenses were granted to specialty services, including two for ethnic, or third-language, broadcasters (CRTC 1984). In the first round of these licenses, Telelatino (or Latinovision as it was known at the time of the initial application) was approved as a joint Italian and Spanish language network, and the other third-language license was granted to a Chinese language service (Chinavision, which would later come to be called Fairchild Television).[5] The move to cable networks with the licenses granted in 1984 marked an effort on the part of regulatory authorities (since both stations were still owned by "independent" parties rather than major media companies) to integrate ethnic media into the national media market.

However, the creation and early development of these two services was hardly successful in its attempt to create sustainable national multicultural television services in Canada. Both Telelatino and Chinavision struggled to cover the costs of operating a cable network.[6] Partially this was due to the difficulties of running a profitable specialty service in the small Canadian market (and was experienced by many of the early services, like the Life Network which went out of business within two years of its creation). However, competition with existing local networks and the new national networks made it particularly difficult to build large enough audiences to make the services sustainable. In the case of Telelatino, there was an explicit attempt to limit competition on the part of the CRTC by forbidding the new channel from competing with local services. In the announcement of their license, it was explicitly stated that the new service would "cooperate with conventional broadcasters and would seek to avoid, wherever possible, any head-to-head programming in the same language" (CRTC 1984). In practice, this effectively blocked the station from producing many kinds of news and information programming, as it would put Telelatino in direct competition with channels like CFMT in Toronto that anchored much of their programming around different third-language news broadcasts.

Along similar lines, Chinavision was barred from broadcasting in the Vancouver market for the first two years of its license out of concern that competition with west coast–based Chinese-language service Cathay Television would be unsustainable (CRTC 1984). After the launch of the services, both channels also had trouble convincing cable companies to pick

---

5 See CRTC (1984), Decision 84-444 (May 24) and CRTC (1984), Decision 84-445 (May 24), for the outcomes of the Latinovision and Chinavision applications.
6 For example, a takeover bid launched in 1988 by Dan Iannuzzi at WTM Media involved disclosing to the CRTC that Telelatino was losing money every year. See Partridge (1988a).

up the service as either part of extended cable packages or on a subscription basis. Often this was because of concerns on the part of cable carriers that the new services would reduce already existing services they offered. Indeed, the CRTC went as far as reprimanding cable carrier Videotron for refusing to offer the services because it felt it was already serving this constituency by carrying Télévision Ethnique du Québec (Partridge 1988b). As a result, both channels incurred significant debt and changed their ownership within ten years of their creation. In 1990, Telelatino restructured its ownership, reducing the majority share owned by founder Emilio Mascia to 51 per cent with the remaining 49 per cent being divided between prominent Italian investors, including the president of several food product companies as well as local Italian broadcaster Johnny Lombardi and the publisher/broadcaster Dan Iannuzzi (Gorrie 1990). Similarly, Chinavision entered into receivership in 1992 with revelations that the owner, Francis Cheung, owed more than $7 million, requiring a new owner (the Fairchild Group) to take over the station.

In order to resolve what seemed to be the systemic problems of operating national ethnic television networks in Canada, two issues emerged during the first decade and half of their existence: repeated attempts to have the requirements for Canadian content lowered and a series of proposals to consolidate the language markets around single broadcasters. With regard to the first issue, Telelatino in particular struggled with meeting the Canadian content requirements outlined in its broadcast license.[7] Contributing to this shortcoming was the fact that the license forbade the production of news programming, usually a low-cost way of meeting Canadian content regulations. As a result, the station was reprimanded several times by the CRTC for failing to meet what were already reduced Canadian content regulations (which were set exceptionally at 15 per cent rather than the normal 40 per cent). When the station met with the Commission for its first license renewal in 1988, the network was granted only a five-month extension on condition that it would make significant changes in the amount of Canadian content it showed and the level of investment in original programming ("Telelatino Gets a ..." 1988). Telelatino argued that only by lowering the required level of Canadian content would it be possible for the station to survive, because then it would be able to lower its operating costs.

At the same time as these discussions about Canadian content were

---

7 This goes back to the earliest days of the network's existence with requests being filed twice within the first three years of Telelatino's launch. See "CRTC Defers Decision on Content" (1985); Enchin (1986).

taking place, there were a number of proposals involving both services to involve them in plans to consolidate the media market. The transformation of Chinavision into Fairchild TV in 1993, taking place after its purchase by the Vancouver-based Fairchild Group, entailed the more or less simultaneous purchase of Cathay International television and their merger into a single Chinese-language network that would be available from coast to coast ("Fung Acquires TV" 1994). Similarly, Telelatino was subject to many proposals that would have involved the consolidation of Italian-language media in Canada. Although none of these deals were ever realized, this involved proposals by Dan Iannuzzi (who later became one of seven shareholders in 1990) in conjunction with Rogers Television ("CRTC Rejects Rogers Unit's Plan…" 1989). Both were rejected because of concerns that this would affect the diversity of offerings available to the viewer. The station was eventually purchased by Corus Entertainment, at which time Mascia sold his remaining stake in the channel.

The development of these two services and their subsequent evolution give some insight into ways that the transition to national third-language television involved an attempt to change the scale at which multiculturalism was regulated. The CRTC was quite forceful in its program to create national services, disciplining cable companies that stood in the way of this goal. It was, however, only partially successful given that the existing industry dynamics and regulatory philosophy prevented the displacement of existing services. When they weren't actively blocking these services from competing with channels of their own, cable providers essentially treated these channels as cable-based local services, limiting their distribution to markets centred around particular urban centres in particular regions. Throughout the 1980s and 1990s, the number of Canadians who were able to see these channels was comparatively low. Stuck between the mandate of the government and market pressures, these channels were national only on paper since they continued to serve the same archipelago of communities that older ethnic services had.

From National Networks to International Services

The second moment of transition in the development of ethnic television in Canada once more coincided with the expansion of carrying capacity on distribution platforms, this time resulting from the transition to digital distribution. This resulted in hundreds of services being approved for distribution in the Canadian market. It was part of the transition between the second and third stage of ethnic broadcasting that witnessed the licensing of dozens of third-language international services. This period saw the transition between two different understandings of broadcast licensing and its relationship to spectrum scarcity. The development of Telelatino and Fairchild TV were part of an approach to ethnic media that essentially sought to replace international services with Canadian-based offerings. The arrival of digital distribution turned away from creating Canadian services toward licensing international services (sometimes in partnership with Canadian media companies.) This effectively moved the scale of regulating third-language and ethnic media away from attempts to structure competition within the national market toward a situation in which several services targeting the same linguistic or cultural market were made available to Canadian viewers, albeit usually on a subscription basis.

Like most Canadian networks, much of the content on Fairchild and Telelatino came from international sources. Fairchild was transformed into what was essentially the Canadian franchise of the Hong Kong-based TVB (short for Television Broadcasts Limited) company in a relationship that went back to the financing for the purchase of Chinavision and Cathay International in the mid-1990s by Thomas Fung (Williamson 1994). This relationship has proven successful and allowed the service to introduce new channels focusing on other Chinese dialects as well as a few other Asian languages. Telelatino, on the other hand, found itself in a much more precarious situation. The network's Italian content was taken from the international services of the Italian public broadcaster, Radiotelevisione Italiana (RAI). While some of this was taken from the 700 annual hours that were provided free to broadcasters around the globe (roughly two hours a day) as part of the public service mandate of the Italian broadcaster, an additional agreement was signed that allowed the Canadian broadcaster to supplement these hours with more programming at a fixed cost that targeted sports programming in particular (Hayward 2009). TVB's investment in Fairchild has meant that the service has avoided direct competition with the Canadian network; however, the transition to digital cable in Canada made agreements like the one between Telelatino and RAI, which were essentially the only way for third-language programming to enter Canada, seem increasingly anachronistic.

Rather than substituting international services with national ones, what emerged during this period was a series of agreements in which Canadian companies (often cable or satellite distributors themselves rather than broadcasters) would become nominal partners of foreign broadcasters in order to get them across the threshold of the Canadian market (CRTC 2000). By 2005, the commission had outlined an exemption for third-language subscription services that made their addition to digital programming packages easier (CRTC 2005). Unlike the previous kinds of ethnic television in Canada, the licenses issued for these new services included minimal or no requirements to produce local or national programming, and the requirement to broadcast Canadian content was reduced from 40 per cent of programming to 15 per cent (the same ratio at which it had been set when the national services were created).

If issues of the relationship between national and local services had been the defining debate during the previous period in the development of third-language television in Canada, the debate shifted with the rise of these new channels, speaking to the new scale at which multicultural media now needed to be regulated. These services were increasingly seen as one of the ways in which the Canadian media market was connected to the international television industry. No longer an issue internal to the nation, these new television offerings were now seen as straddling the increasingly porous and volatile borders of the nation itself. Granted, these are concerns that stretch back to the earliest days of ethnic media in North America (and, at the time of its initial proposal to purchase the failing Chinavision, there were concerns that the Fairchild Group would serve as an instrument of the Chinese government). However, these became quite pointed during in the early part of the twenty-first century (Rusk 1993).

The new scale of ethnic media in Canada became most apparent in the debates that surrounded the introduction of the Italian-language RAI International to the Canadian market. This was a debate that also involved questions over the possible instrumentalization of television for ethnic communities more broadly, most notably in the case of the Al-Jazeera news network. In the case of RAI International's application, it put the station in direct competition with Telelatino. Also, since RAI was one of the older network's primary sources for programming, it presented a serious threat to the business model of the station. The details of this episode have been outlined in greater detail elsewhere; for the purposes of this chapter it is sufficient to draw out only a few aspects of the ensuing conflict between RAI International and Telelatino. Most important among these is the degree to which the rhetoric that accompanied RAI International's application made much of the importance the service would have for the electoral rights of Italian citizens residing in Canada (many of them having been naturalized in the three decades since the last wave of migration from Italy in the 1970s). The introduction of the new television service was intimately linked to the globalization of electoral democracy in Italy. The debates that followed involved not just the broadcasters involved, the commission, and interested stakeholders; it also involved the Department of Foreign Affairs and Trade (DFAIT) and the cabinet at the time. It was a debate in which older, nationally framed understandings of multicultural-ism and assimilation came into conflict with an understanding of cultural diversity shaped by the era of digital technology and expanded global mobility.[8]

Eventually, RAI International was granted access to the Canadian mar-ket. The years that followed saw the development of a much more clearly articulated framework for the licensing of third-language channels. The decision to license RAI International was one of two test cases (Al-Jazeera being the other) in the development of the new policy and involved a considerable amount of pressure from many constituencies both in favour and in opposition to the application. In the years between 2004 and 2006, the commission laid out the criteria for licensing non-Canadian, ethnic services (CRTC 2005). It was significantly easier for new third-language services more explicitly, now referred to as international channels by cable companies, to enter the Canadian market. Along with the already men-tioned reduction in Canadian content requirements, the policy sought to limit direct competition with existing services by requiring cable provid-ers interested in launching particular services broadcasting more than 40 per cent of their content in one of the languages already covered by an existing service to offer the older service to its viewers as well. Calling this a "buy-through," it was an attempt to ensure that cable companies did not abandon the national ethnic networks for the new services, many of which they had sponsored for entry into the country.

Over the past decade, there has been a significant (and rapid) increase in the number of third-language television networks available in Canada. While it is perhaps too soon to comment on the long-term consequences, it is worth noting how both Telelatino and Fairchild TV have adapted to the changing dynamics of television in Canada. As part of a larger corpor-ate group, Fairchild has expanded its services to include both Cantonese and Mandarin channels as well as some programming in Korean and Vietnamese. These are spread across two channels, Fairchild TV and Talentvision, and the company asked for two more channels (known as Fairchild II and Talentvision II), although these have yet to be launched. It still broadcasts almost all of its content in one of the four languages mentioned. This development is certainly also a product of the population growth and maturation of Chinese-Canadian communities into sizable,

---

8 A similar situation was seen with regard to the licensing of Al-Jazeera, where concerns over possibly inflammatory programming involved not just the CRTC, but also direct intervention from the DFAIT.

wealthy audiences in many parts of the country. Telelatino, on the other hand, has adapted to the changing demographics of Italians in Canada and increasingly moved toward ethnic, rather than third-language, programming. This has involved repeatedly asking permission to broadcast more English-language programming as well as broadcasting American content (such as *The Sopranos*) translated into Italian and Spanish. The station has also come to develop the profitability of its Spanish-language programming, as the hispanophone community in Canada has grown and become wealthier. While both channels are considerably more financially secure than in their early days, their development speaks to the way in which the fate of third-language and ethnic television is entwined with the changing cultural dynamics of the nation itself.

At the level of regulation, however, the introduction of digital distribution raised many issues about the ways in which the national television industry is regulated with regard to the dynamics of international media. These were issues which, while of significance in the earliest days of media, had mostly been put to rest during much of the twentieth century when national communications policies could be framed in relation to settled borders. The commission takes an increasingly light-handed approach with regard to reducing and regulating competition, but decisions regarding how to license these services increasingly entail engagements with the complexities of international politics. It would not be surprising if, in the years to come, the DFAIT plays an increasingly visible role in managing the development of cultural diversity in Canadian media. This speaks directly to the emergent scale of media policy, a development which involves audiences that are both increasingly fragmented and focused in their consumption as well as being increasingly connected to other similar audiences around the world.

Conclusion

There are many narratives that might be offered to make sense of the development and transformation of Canadian multicultural television. These range from discussions of the expanding diversity of offerings available to Canadian television viewers that came with the transition from terrestrial to cable to digital modes of distribution, to descriptions of the mainstreaming of multicultural media (a trend accompanied by large media companies taking over these services). However, the goal of discussing these two periods of transition has been to highlight the complex interaction between policies supporting cultural diversity, developments in the organization of the Canadian television industry, and the changing scales of media governance over the past three decades. This is to say, it is not simply that any one of these factors individually captures the evolution of multicultural television, but rather the interaction between all of these trends. The way in which these developments have taken shape speaks to what has been described in this chapter as the changing scale on which the television industry operates and is regulated as well as how identity and cultural diversity are imagined.

To speak in general terms: the movement from the "local" to the "national"—the first period of transition discussed here—was involved in a much broader attempt to reconfigure existing institutions, identities, and circuits of distribution that linked ethnic communities to their countries of origin and adoption so that they would coincide with national boundaries. Granted, the specific community dynamics of Italian-, Spanish-, and Chinese-speaking Canadians are quite different from each other. However, the goal of the policy can broadly be seen as an attempt to short-circuit (or at least inflect with a Canadian accent) those information networks that linked community members to institutions and information sources outside of Canada. This is clear from the ongoing discussions, reprimands, and re-calibrations that these services engaged in regarding the percentage of Canadian content they were required to show. As for the movement from the "national" to the "international" that characterized the second moment of transition discussed above, it is marked by a turn away from many of the most important aspects of the policies that created "national" services in the 1980s, including the CRTC's absolute preference for Canadian services over international services and the practice regarding direct competition between similar services (a trend that can now be seen as common across the industry). This transition can be seen as a part of the reintegration of Canadian media into international markets, a common global trend during these years.

There is a longer discussion that would be warranted about the way in which these transformations in media policy also speak to the changing contours of the politics of cultural identity in Canada. It has sometimes been noted that the creation of both Telelatino and Chinavision were deeply influenced by the desire on the part of the Trudeau-era Liberal Party to maintain its relationship with ethnic communities across the nation (Roth 1998). The move to digital distribution platforms, and the expanded space made available for third-language services after 2000, speaks directly to the growing influence of a wide variety of communities across the country. At the same time, however, it also gave rise to an increasingly nativist voice concerned about excessive foreign influence over affairs internal to Canadian society. This was the case with the debate over whether or not to license RAI International in Canada as well as anxieties about the introduction of Al-Jazeera to the Canadian market. While most Canadian politicians are hesitant to declare the death of multiculturalism (unlike some European leaders), there can be no doubt that the institutional structures and discursive tenor that shape how people speak about cultural diversity has shifted. Multiculturalism in Canada today is both more open and more conflictual, and these are trends that are borne out with regard to television and other communications media. However, perhaps this marks not so much the "end" of multiculturalism (and certainly not its death), but its maturation as a stable yet constantly shifting component of Canadian culture.

References

Amatiello, Michael. 2009. "CHIN Radio: A New Canadian's Guide to Good Citizenship." MA thesis, York University.

Beatty, Bart, and Rebecca Sullivan. 2006. *Canadian Television Today*. Calgary: University of Calgary Press.

Bissoondath, Neil. 1994. *Selling Illusions: The Cult of Multiculturalism in Canada*. Toronto: Penguin Books.

Canada. Royal Commission on Bilingualism and Biculturalism (RCBB). 1969. *Report of The Royal Commission on Bilingualism and Biculturalism*. 5 books. Ottawa: Queen's Printer.

Canada. Royal Commission on Biculturalism and Bilingualism. 1969. *The Cultural Contribution of the Other Ethnic Groups*, Book 4. Ottawa: Queen's Printer.

Canadian Radio-television and Telecommunications Commission (CRTC). 1983. "Call for new Specialty Programming Services." CRTC Public Notice 1983-93 (4 May).

———. 1984. Decision 84-444 (May 24). Accessed June 14, 2012. http://www.crtc.gc.ca/eng/archive/1984/DB84-444.htm

———. 1984. Decision 84-445 (May 24). Accessed June 14, 2012. http://www.crtc.gc.ca/eng/archive/1984/DB84-445.htm

———. 1985. "A Broadcasting Policy Reflecting Canada's Linguistic and Cultural Diversity." *CRTC Public Notice 1985-139* (July 4). Accessed June 14, 2012. http://www.crtc.gc.ca/eng/archive/1985/PB85-139.HTM

———. 1986. Decision 86-586 (June 19). Accessed June 14, 2012. http://www.crtc.gc.ca/eng/archive/1986/DB86-586.htm

———. 2000. "Licensing of Digital Pay and Specialty Services." *CRTC Public Notice 200-171* (December 14). Accessed June 14, 2012. http://www.crtc.gc.ca/eng/archive/2000/pb2000-171.htm

———. 2005. "Revised Approach for the Consideration of Broadcasting Licence Applications Proposing New Third-Language Ethnic Category 2 Pay and Specialty Services." *CRTC Public Notice 2005-104* (November 23). Accessed June 14, 2012. http://www.crtc.gc.ca/eng/archive/2005/pb2005-104.htm.

"CRTC Rejects Rogers Unit's Plan To Buy Telelatino." 1989. *Globe and Mail*, July 28, B5.

Cavanaugh, Richard. 2004. *Reflecting Canadian: Best Practices for Cultural Diversity in Private Television*. Toronto: Canadian Association of Broadcasters.

Collins, Richard. 1990. *Culture, Communication & National Identity*. Toronto: University of Toronto Press.

Day, Richard J.F. 2000. *Multiculturalism and the History of Canadian Diversity*. Toronto: University of Toronto Press.

Edwardson, Ryan. 2004. "Other Canadian Voices: The development of Ethnic Broadcasting in Canada."Iin *Racism, Eh? A Critical Inter-disciplinary anthology of Race and Racism in Canada*. ed. Charmaine Nelson and Camille Antoinette Nelson, 316–325. Toronto: Captus Press.

———. 2008. *Canadian Content: Culture and the Quest for Nationhood*. Toronto: University of Toronto Press.

Enchin, Harvey. "CRTC Defers Decision on Content." 1985. *Globe and Mail*, July 23, B16.

——. 1986. "CRTC to Consider Bids to Amend Content Rules." *Globe and Mail*, February 15, B4.

"Fung Acquires TV." 1994. *Advertising Age*, March 29, I-6.

Gorrie, Peter. 1990. "Six Allowed to Buy 51% of Telelatino." *Toronto Star*, July 10, F3.

Hayward, Mark. 2009. "'Il Caso Canadese' and the Question of Global Media." In *Beyond Monopoly*, ed. M. Ardizzoni and C. Ferrari, 21–35. Lanham: Lexington Press.

Hirji, Faiza, and Karim H. Karim, eds. 2009. "Race, Ethnicity, and Intercultural Communication." Special Issue, *Canadian Journal of Communication* 34 (4).

Horsti, Karina. 2009. "From Multiculturalism to Discourse of Cultural Diversity: European Public Service Broadcasting and Challenge of Migration." Paper presented at ???, Lisbon. Accessed November 1, 2011. http://imiscoecrosscluster.weebly.com/uploads/4/6/9/4/469440/lisbon_paper2.pdf

Karim, Karim H. 2009. "Other Research Reports on Issues of Race, Ethnicity, and Communication." *Canadian Journal of Communication* 34 (4): 749–755.

Kymlicka, Will. 2001. *Finding Our Way: Rethinking Ethnocultural Relations in Canada*. Toronto: Oxford University Press.

Partridge, John. 1988a. "Tough Review in the Offing for Pay TV." *Globe and Mail*, June 13, B1.

——. 1988b. "Videotron rapped for Usurping Powers of CRTC." *Globe and Mail*, June 21, B5.

Raboy, Marc. 1990. *Missed Opportunities: The Story of Canada's Broadcasting Policy*. Montreal: McGill-Queen's University Press.

Roth, Lorna. 1998. "The Delicate Acts of 'Colour Balancing': Multiculturalism and Canadian Television Broadcasting Policies and Practices." *Canadian Journal of Communication* 23 (4): Available at: <http://www.cjc-online.ca/index.php/journal/article/view/1061/967>. Date accessed: 14 Jun. 2012.

——. 2005. *Something New in the Air: The Story of First Peoples Television Broadcasting in Canada* Montreal: McGill-Queen's University Press, 2005.

Rusk, James. 1993. "Beijing Interference Feared if Chinavision Sale Proceeds." *Globe and Mail*, July 3, A7.

Stein, Janet Gross. 2007. *Uneasy Partners: Multiculturalism and Rights in Canada*. Waterloo, ON: Wilfrid Laurier University Press.

Stone, Phil. 1976. "You Know Johnny Lombardi?" *Broadcaster*, June, 6.

Taylor, Charles. 1994. *Multiculturalism: Examining the Politics of Recognition*. Princeton: Princeton University Press.

"Telelatino Gets a Reprieve from CRTC." 1988. *Globe and Mail*, December 23, B2.

Williamson, Robert. 1994. "Media Baron Rides Human Wave." *Globe and Mail*, February 19, B1.

12

Old Media, New Media, Intermedia: The *Toronto Star* and CFCA, 1922–1933

Sandra Gabriele and Paul S. Moore

Early radio was *read* and *seen* as much as it was listened to. In fact, early radio audiences were inextricable from newspaper readerships. The link goes beyond cross promotion and publicity, but the most direct connection is the best starting point: newspapers owned many of the first radio broadcast licenses in Canada and built some of the first commercial radio stations in North America. Leading the way in Ontario, the *Toronto Star* began broadcasting through "wireless telephone" on March 28, 1922. Its first broadcast did not feature breaking news, weather, financial, or agricultural market reports. Instead, the broadcast was a musical concert featuring local singers and instrumentalists providing a selection of classical music, and the "syncopated harmony" of jazz. The performance was transmitted from a remote studio to an estimated one thousand homes already equipped with amateur receiving sets. Two special audiences also gathered to listen together: the general public at the Masonic Temple and First World War veterans at the Christie Street Military Hospital. The front page coverage the following day extolled not the quality of the musical performances, but the clarity of the broadcasts, with careful note of the range of the received broadcast for the benefit of amateur "radio fans" scattered more than a hundred miles beyond the city limits. From Belleville, Ontario, to Buffalo, New York, radio fans wrote letters praising the technical quality of the broadcast (March 29, 1922, 1).

In a time when the "wonder" of radio was itself a spectacle, the *Star*'s coverage of the event focused almost exclusively on the technical element of broadcasting, asking readers with radio receiving stations at home to share not only the quality of their reception, but also to specify what kind

218

of set they were using. In these early days of experimentation with radio, broadcasting was not simply a novelty entertainment; it was a technological wonder that "marked a new step forward in Canadian radio circles" (March 29, 1922, 1). The desire of any popular newspaper was not simply to cover, but to be a *part* of such a modern, technological spectacle. In its editorial following the inaugural radio concert, the *Star*'s editors, like all proponents of radio's potential, imagined the "limitless" possibilities forged by the wireless, global reach of this new technology, "transcending imagination" (March 29, 1922, 6). The latent message of the editorial suggests that newspapers valued radio because it could transcend the temporal and spatial limitations of circulating newsprint. The *Star* continued the musical concert series weekly, and soon began a nightly thirty-minute program (April 10, 1922, 1). While the concert series, like the first broadcast, was entirely musical, the nightly program was more varied, consisting of music, topical lectures, a children's bedtime story, and sporting and financial bulletins.[1] With the completion of the *Toronto Star*'s own transmitting station, CFCA, atop its head office building, the nightly program was expanded to one hour in June 1922, but its variety format remained the same balance of music, stories, business and sports bulletins, and brief feature lectures (June 23, 1922, 10, 13).[2] Despite the integration of the studio within the *Star*'s newspaper office, it took another few months for CFCA to include the *Star*'s news content in its radio programs.

Pervasive synergies of cross-media ownership since the introduction of radio in the 1920s make direct investments across media commonplace. Yet there was nothing inevitable about newspapers such as the *Toronto Star* staking a claim to the centre of early radio. Why would a newspaper venture into broadcasting? And why would it begin with musical programming to a public audience instead of a newscast to home listeners? This chapter sets out to answer these questions, placing the case study of the *Toronto Star*'s CFCA within the context of a broadening role for newspapers in an emerging mass popular culture at the turn of the twentieth century. The combined provision of leisure and cultural technologies as supplements to news and information established an expansive social function for newspapers in both their printed form and their publishing companies. In that context,

---

1 The *Star*'s first few daily broadcasts also included a telegraphed report from its parliamentary correspondent in Ottawa, a feature quickly dropped from the variety format.

2 In the first years of radio, a licensed "station" and its call letters referred to the equipment and location of a specific transmitter, rather than a frequency on the radio spectrum. Initially, while the *Star* awaited the construction of its own "station," its programs were transmitted from 9AH (later CKCE) in co-operation with the electronics manufacturer Canadian Independent Telephone Company.

radio could indeed be understood beyond the confines of a promotional or business venture as an innovative feature of the newspaper form itself.

The Newspaper Supplement and Intermediality

Newspaper-owned radio stations did not dominate the new broadcasting industry in Canada for very long. In fact, their brief importance predates the policy preoccupations of Canadian nationalism that emerged as a result of commercialized, American network radio later in the 1920s (Dorland 1996, xi; specific to radio, see Jackson 1999).[3] Studies of Canadian cultural industries, including earlier editions of this very book, have largely been shaped by this "policy reflex" (Wagman 2010). This is especially true of radio, although recent scholarship has turned to audience reception and listening cultures, alongside continued interest in representation and programming.[4] Our focus on the beginnings of radio broadcasting in Canada, when newspaper-owned stations predominated, turns to the public mediation of an increasingly integrated media culture in order to better understand the mutually constitutive nature of these media forms. Radio listening took shape, and became a meaningful cultural practice, through newspaper reading, making both audiences one media public. The newspaper organized its radio programs to fulfill an experience begun by reading its printed stories, and in turn edited its print columns (especially its radio page) as an essential component of listening to its broadcasting.

This case study paints in details at the conceptual "edges" of the technologies of radio and newsprint by looking at the moment when the intermedial relation is most overtly at play (Marvin 1988, 8). Éric Méchoulan (2003) has argued that intermediality "functions on the level of the media melting pot from which a well-defined medium slowly emerges and is institutionalized" (22). In other words, intermediality is a conceptual tool for describing that which "precedes the medium" as a recognizable, distinct technology.[5] Although most evident with the introduction of a new

---

3 In Canada, as is well known, debates over commercialization and public interest led to a Royal Commission on Radio Broadcasting in 1929, the nationalized radio programming of the Canadian Radio Broadcasting Commission in 1932, and the nationalized radio network of the Canadian Broadcasting Corporation in 1936 (Vipond 1992; Raboy 1990). These policy interventions did not upend the dominance of commercial companies relying upon popular content produced in the United States, even on the CBC itself (MacLennan 2005; Skinner 2005; Vipond 1999).

4 Among a growing body of research on Canadian radio history, we draw attention to MacLennan (2008, 2012), Vipond (2009, 2010), and Kuffert (2009) for their studies of early listening cultures, as well as Webb (2008) for his social history of a regional broadcaster.

5 We borrow our understanding of media from Lisa Gitelman (2006), who argues media are socially realized structures of communication, including the hardware and social protocols necessary for understanding it and the cultural practices that make up the communication process (7). The

medium, the intermedial relation between newspaper and radio in this case is supplementary, not just complementary. The character and cultural importance of the one medium is incomplete—even meaningless—without the encounter with the other. Rather than privilege radio in its relation to the newspaper by defining the case study as a genealogy of "listening in," we are more broadly interested in the ways intermediality operates through a singular mode of address, with news serving as part of the variety of modern leisure. In exploring the propensity for newspapers to be early leaders in owning and operating radio stations, we focus specifically on their capacity and expertise in providing variety, leisure, and amusement as supplements to the news. As a result of this approach, we trace the exhibition and visual display of radio listening—its illustrated logic—as practices borrowed from newspaper supplements. While this methodology derives from the necessary constraint of turning to archives of print documents, not least the newspaper itself, we demonstrate nonetheless that the act of listening to radio in its earliest years was constantly supported by looking, writing, reading, depicting, and visually documenting the novelty of the practice. Both at the time and looking back now, these practices are all the more important because of radio's ephemeral, ethereal qualities.

The North American newspaper became a nexus of popular culture at the turn of the twentieth century. Central to this transformation was the incorporation of expansive, illustrated weekend supplements (Barnhurst and Nerone 2001; Rutherford 1982). The weekend supplement included extensive commentary, directories, and advertising for all sorts of popular amusements and consumer technologies, but its material form was even more important. Through the weekend supplement, newspapers drew upon aspects of other media, including serialized novels, magazines, posters, and sheet music. Different qualities and sizes of paper and colour printing encouraged readers to collect and treat parts of the newspaper as keepsakes. Weekend supplements also took a central place by employing new media forms such as comic strips and serialized moving pictures.[6] As we have argued elsewhere (Gabriele and Moore 2009), Canadian newspapers developed such features in tandem with their counterparts in the

---

development of those protocols that eventually made each medium distinct from the others was a social and cultural process. We argue that the process for radio was established through print culture. Further, however, the practices of listening were also essential for developing a literacy for understanding the print radio columns. Our focus here is on newspapers, though it took place in other print media as well (Douglas 1987, 1999).

6 For the *Toronto Star* in particular, features such as illustrated and colour comic supplements appeared in the *Star Weekly* after its creation in 1910. Archived and microfilmed separately from the daily *Star*, and in later decades taking the form of a national magazine, the *Star Weekly* is often neglected despite functioning in the 1910s and 1920s as an expanded weekend edition.

United States, although following a path of nationalist distinction with a civic emphasis on informativeness and educational value against the sensationalist, secular metropolitanism of the "American Sunday paper." In the first decade of the twentieth century, American newspaper syndicates standardized and disseminated much of this material, so that weekend supplements such as magazines and comic strips entered the daily press of smaller cities, including in Canada, and were no longer the exclusive domain of metropolitan newspapers in the United States.

While newspapers' primary identity and business model remained rooted in print sales, many of their weekend supplements experimented with intermedial forms, inserting tabloid novels, song sheets, and glossy magazines, and even collaborating to produce newsreels. By 1922, newspapers were practiced in adopting such intermedial strategies for expanding circulation, and newspaper reading had already come to include other activities. The paper itself now included cut-out toys, puzzles, and collectible inserts; its amusement and sporting columns helped schedule leisure time, as its advertising helped schedule and direct consumption. Further, newspapers frequently sponsored community events, from trade shows to charity concerts to election night gatherings that turned reports of voting returns into popular festivals. While this is clearly a matter of commercialization, the emergence of weekend supplements as a central part of popular culture and a significant part of newspaper publishing should be taken as an extension of publicity. In the sense of making public, or constituting a public, publicity transforms the pursuit of increased circulation into a complex relation of public service and popular appeal, not reducible to maximizing advertising and profits. This is not to deny the importance of circulation as a matter of sales, but rather to attenuate that aspect as only part of the social relation of publicity. In 1922, newspaper readers were addressed as a social and civic public, not just a market. The relationship between newspapers and their newly formed radio stations was similarly not simply commercial. Newspapers understood radio as enabling a fulfillment of their mandate of public service, including, but not limited to, providing popular entertainment in the mode of weekend supplements. The connection was explicit for Judith Walker, who ran the *Chicago Daily News* station WMAQ: "When I thought of a women's program, I would think of it emanating from the women's department of the paper...We tried to tie the paper and the station together" (cited in Hilmes 1997, 72).

Radio as an Extension of the Public Service of Newspapers

Newspaper publishers owned a significant proportion of the earliest radio stations in Canada, as in the United States.[7] About half of the first two dozen commercial broadcasting licenses in Canada were issued to newspapers.[8] Scholarly explanations of the early adoption of radio by newspapers build upon the contemporary rationales given by newspaper publishers themselves. First is the instrumental logic of cross promotion in the face of competition from other newspapers (Allen 2009, 51). Owning a radio station meant broadcasting the newspaper name as a form of advertising to encourage buying the paper itself.

A second explanation is that newspapers were keen to take up fads and novelties, or phrased more charitably, to adopt new technologies in order to fulfill their established place in the cultural vanguard. In Montreal, *La Presse* already had plans for one of the most powerful radio transmitters in North America when it began a radio column to keep its readers informed about "la grande merveille de la téléphonie sans fil" (April 29, 1922, 1). Finally, newspaper publishers were also likely to espouse a principle of public service, though this sentiment could be self-serving in and of itself. For the manager of the *Kansas City Star*'s WDAF, "in terms of management, public service, and program quality, the newspaper was simply the best kind of owner" (cited in Stamm 2011, 14).

Mary Vipond (1992) concisely proposes all three factors in her comprehensive history of the first decade of radio broadcasting in Canada: newspapers perceived an affinity with radio as the latest fad to secure publicity and therefore sales, but also took the more principled stance of radio serving "as a natural extension of and supplement to their role as media of communication" (44). Announcing with great fanfare its "free wireless and radiophone service," the *Vancouver Sun*'s "advanced method of news distribution" was just one of the latest technologies used to provide "the best newspaper service to the public that science permits" (March 7, 1922, 1). Radio would thus allow the *Sun* "to contribute its quota in the

---

7 Compared to Canada, it is important to note a reversed trajectory of newspaper-ownership of radio stations in the United States, which begin to dominate throughout the 1930s, whereas Canadian newspapers largely withdraw from broadcasting with its nationalization in Canada in the 1930s. According to Stamm (2011), newspapers in the US in the 1930s pioneered the techniques of media convergence precisely through operating radio stations, a trend that "continued into newly licensed FM stations after 1941 and television stations after 1948" (5).

8 Newspapers owned ten of twenty-three early radio broadcasting licenses listed in the *Winnipeg Tribune* (April 25, 1922, 5), and four more were owned by George Melrose Bell, publisher of the *Regina Leader*, which assumed one of his licenses upon opening CKCK in Regina. Vipond (1992, 21), citing a different source, lists twenty-one of these, with both discrepancies newspaper-owned: the *Calgary Herald* CHCQ and *Edmonton Journal* CJCA.

interest of commercial and social progress...The undertaking will prove to be the most popular feature ever adopted by any newspaper" (March 9, 1922, 1). In hindsight, positioning radio as a "feature" of the newspaper seems incongruous because radio developed, in time, its own form and distinct identity; but conflating the newspaper-radio program and the radio-newspaper page regularly occurred within the press. When *La Presse* solicited print advertisers nationally, it called its radio page "Canada's Greatest Radio Medium," and focused on details about its radio station rather than the newspaper's circulation (Toronto *Globe*, May 13, 1922, 7). The interchangeable address of readers and listeners is a key logic behind how the *Toronto Star* edited its radio page and programmed its broadcasts.

Like other newspapers of the time, the *Star* was an avid publicist of the newest technologies (cf. Dooley 2007). Its interest in radio preceded its entry into broadcasting, in much the same way that it rhapsodized about other technologies.[9] It was not surprising that during the inaugural broadcast, the *Star*'s managing editor, J.R. Bone, would reference other technologies in considering how experimenting with radio was part of the public service that the *Star* provided its readers: "The *Star*'s interest is simply that...its primary duty is to give its readers the news...the most practicable way of dealing with it from a news point of view is not only to describe what is being done by others, *but to give practical and public demonstrations of the invention*...It is hoped, also, that the service which the *Star* will send out hereafter will be a real service to the city and province in supplying information and entertainment" (March 29, 1922, 2; emphasis added). Bone's speech reveals the rationale for the relationship that the newspaper imagined it would have to the radio station and that both would have to the audience. Keeping its readers abreast of the latest news required the newspaper not simply to report, but to do: to take part in the development of radio's potential through experimentation. Upon the closing of CFCA in 1933, this logic was explicitly stated by the paper: "[The *Toronto Star*] in its whole history has always been alert on behalf of the public for the new and significant, and which *does things* as well as writes about them when they are done" (August 29, 1933, 6; emphasis added).

Public service also required publicity, for the newspaper embodied both information and entertainment. The *Star* led the way in the Toronto market for promotional schemes, designed to increase circulation by promoting the paper itself. The job of doing this fell to the Promotions Editor,

---

9 For example, the *Star* claimed to be the first newspaper in Canada to use wireless to send location reports to the newsroom (August 10, 1903, 1), and later profiled the technology of its new printing press and composition room (August 26, 1905, 1–7).

William Main Johnson. Johnson began work with the *Star* in 1910 as a reporter; after the First World War, he served as a Parliamentary correspondent in Ottawa. By 1921, Johnson began the specialty editorial work of weekend papers writing "special articles," becoming the pictures editor for both the daily and weekly editions in 1922—a vital position, given the pictorial page's position as the back cover of the daily edition, and the rotogravure section's function as the cover of the *Star Weekly*. Later that same year, this work expanded when he took charge of "editorial promotion" for the *Star*. This work involved crafting and carrying out promotional activities that aimed to cultivate an active and engaged relationship with readers. These schemes positioned the newspaper as a vehicle publicizing planned activities. However, they frequently went beyond the newspaper page, bringing the community together through events that exceeded their typical roles as readers or listeners. Examples of such schemes include free parties for children in local parks with organized games, sports activities, and treats like ice cream cones.

One prominent coupon "voting" scheme involved the purchase of an animal for the Toronto Zoo, as selected by children by sending in ballots marked with their preferences. Elephant "Baby Stella," as she was named by readers, two white peacocks and Tiny Tim, a bear, were eventually purchased throughout 1923. Stella and her companion peacocks were introduced during a twenty-eight-mile-long parade that ran from Oakville to Toronto and included an accompanying orchestra and floats (June 15, 1923, 1). These animals continued to appear throughout the pages of the *Star* over the following year, part of a tactic Johnson called the "accumulative effect," which linked promotional schemes together (letter, Johnson to Ralph Pulitzer, April 28, 1924, 2. Box 2, Correspondence, 1910–1931. File 5, Job Applications & references, 1912, 1924–1931). Such promotional schemes were important because they built certain audience relations, part of the "service" Bone referenced during the inaugural broadcast. In the words of a contemporary working at the Boston *Advertiser*, this form of promotional work carefully included "a tone of friendliness and sincerity that must be highly effective in the making of permanent attachment to the paper" (Edgar D. Shaw to Johnson, May 6, 1924).[10] The attachment to the paper was not only important for the promotional elements that obviously led to greater circulation and popularity, but was also viewed by the paper as part of the way it serviced the community by bringing it together.

---

10 This and the following quotation are taken from William Main Johnson, Toronto Public Library, Fonds L33, Box 2, Correspondence 1910–1931, Folder 5, Job Applications and References, 1912–1931.

Within this context, holding a concert for the inaugural broadcast by CFCA makes perfect sense. Although Johnson described his work as promotions editor as involving "the problem of appealing to the whole mass of the people, the same constituency which a departmental store appeals to" (letter to C.L. Burton of Robert Simpson's Co.; December 19, 1930), children were understood as a specific "medium for this purpose": "this specialized policy of interesting the youngest generation and through them all the other generations has proved a marked success" (Johnson to Pulitzer, April 28, 1924). The interest in child readers specifically coincided with the enthusiasm young boys had for radio experimentation. Indeed, Susan Douglas (1999) notes that young middle-class boys were avid radio fans and marked a special constituency of early radio enthusiasts (65–69). As Hilmes (1997) points out, and the recent work by Elena Razlogova (2011) confirms, early radio enthusiasts were more diverse than Douglas's early work suggests. Indeed, Hilmes notes how the focus in the mainstream press on white, middle-class boys and their shenanigans was strategic fodder for the intervention by "responsible corporations" (39). Nonetheless, boys were frequently addressed within the pages of the newspaper directly, integrating the wider appeal to young readers through the paper's youth section and through its promotional schemes.

Audiences do not merely exist; rather, they are constituted by various kinds of rhetorical addresses. The specific rhetoric of the printed page has been particularly important in this respect (see for example Anderson 1983; Martin 2004; Nord 2001). Michael Warner (2002) notes that texts and their circulation produce a very specific kind of public. Promotional schemes that extended the reach of the newspaper off the page (including radio) were essential in creating a constituent audience for the newspaper (and its radio station) that organized people into reading, and eventually listening, publics. Such publics were always engaged to perform a wide range of tasks (like clipping coupons, or counting words) or gathered to participate in events that went far beyond simply buying or reading a newspaper. Entertaining audiences was an integral part of this process. The focus on children not only ensured a generation of readers to come, but it extended the mass appeal through the domestic setting that was central to the newspaper's mission since the 1890s. The emergence of radio must be seen within the larger context of establishing the family as a key reading public, a way to get the "Whole Human Family United by Wireless" (March 11, 1922, 2). The inter-generational habits of reading were especially relevant for the multiple sections of the *Star Weekly*: its "family" of comic pages, rotogravure cover, serial fiction, and its entertainment, sporting, and automotive columns. The centrality of children for the newspaper's social function, public service, and mass appeal becomes especially important to keep in mind as we turn to CFCA and the promotion of radio.

Radiating Fans and the Mass Public for Newspaper-Radio

A little chap said to him, "Gee, mister, it's a real radio."
"What does it radiate?" asked the MPP, by way of carrying on
*   the conversation.*
"Why, fans, of course," answered the boy, amazed at such
*   inutterable ignorance.*
                                    —"Radio Film Proved Magnet,"
                                    Toronto Star, August 30, 1922, 26

The *Star* declared 1922 "Radio Year" at the Canadian National Exhibition, primarily due to the presence of its own exhibition hall. The Exhibition had long been central to the public adoption of all things modern in Canada (Walden 1997). The *Star's* fairground building featured CFCA over loudspeakers, displays of radio technology, and a special film called *Radio*, co-produced by the local newsreel company Filmcraft. The scene at the *Star's* Radio building was dramatized daily in news items filed from the fair. In these early days of broadcasting, the *Star's* Exhibition building was a key site for making more tangible the ephemerality of radio's radiating waves, and for gathering readers and enthusiasts in one physical location. Beyond being an educational site demonstrating the production of radio, the Radio building was characterized quite literally as producing radio listeners—enthusiastic, dedicated listeners. "Fresh fans are recruited daily from among thousands" was the headline summary for the last days of the Exhibition (September 8, 1922, 10). Newspaper readers were now avid radio "fans."

The transformation of scattered radio amateurs into a mass of radio fans is the overarching discourse in 1922 in the pages of the *Star*. The year began with a Canadian Press dispatch from Ottawa, hinting at the forthcoming regulatory enforcement of government licensing of both sending and receiving wireless stations (January 13, 1922, 1). In that context, the *Star's* daily programming on CFCA would stand paternalistically as a moderating force as well as a mediating institution, creating its audience of radio "fans" by connecting the leagues of young boys on the amateur front with the social pillars of religious, educational, and cultural institutions. For early concerts, "amateurs" with receiving sets at home were asked to "please accept this invitation and co-operate by 'standing by'" during the broadcast, while "radio fans" were encouraged to participate fully in the event by being "good enough to write The *Star* at once" (March 28, 1922, 1). Wireless amateurs (now turned into radio fans) did not disappoint, helping to produce the daily columns of the *Star*'s new Radio Department the following week, sending in letters, commentary, advice, questions, and even photographs of themselves and their tuning outfits. At the same time as the *Star* began its nightly radio program, its printed radio page included a daily comic strip, *Radio Ralf*, by Jack Wilson for the McClure Newspaper Syndicate, whose protagonist was a typical kid in the precocious comic mode. A small story of a real, Toronto "Ralf" was printed on the very same page: "The first thing Douglas W. Sparling and his friends do every afternoon when The *Star* arrives is turn to the Radio page." (April 12, 1922, 10). While the young male radio fan was key to "tuning in" the whole family, the nightly variety provided something for everyone: sports and business reports, classical and popular music, informative lectures, and a bedtime story.

While *Radio Ralf* aimed to mirror the initial importance of juvenile boys as prototypical radio fans, a different daily comic strip came to the radio page in July, bringing with it a new gimmick more directly tied to the wide appeal of the variety on the nightly radio program. Created by Al Posen in 1921, United Feature Syndicate's *Them Days Is Gone Forever* was already running in dozens of newspapers across North America, providing something akin to an illustrated song (Altman 2004, 182–192). The *Star* placed it amidst the features of the radio page. Each day's strip was a rhyming lyric of a comedic song, complete with a line of musical notes to sing along. While other newspapers promoted "Them Days" for the ability to "sing a comic strip" (*Oakland Tribune*, August 13, 1922) and read a "musical cartoon" (*Baltimore Sun*, June 11, 1922), the *Star* transformed it by actually broadcasting the lyrical tune on its nightly radio show using local musicians. Theatrical musicals and song sheet adaptations had long exploited the popularity of newspapers' comic characters (Gordon 1998, 80–81). Now, with radio, the intermedial play of a serial mix of comic strip and comic song happened just hours apart. The co-ordinated variety of the radio page and the radio program was most overt with the "Them Days" comic song, but also used were features such as letters from fans about their listening experiences, pictures of their radio sets, answers to listeners' written questions, not to mention listing the next radio program's lineup alongside photos of featured singers and musicians. Such explicitly intermedial relations between the radio program and the radio page make it evident that they were addressing a single public in tandem.

With its own CFCA transmitter fully functioning atop the *Star's* office building in June, tangential radio ventures began to supplement the work of broadcasting. The *Star's* Radio Car was one such project—one that further coordinated the paper's address to its audiences across a range of media. In July 1922, the *Star* built the travelling radio receiving station to provide CFCA radio concerts for remote audiences at parks across its entire listening radius, throughout the city and beyond, to "give the public a more complete radio service...you may see and hear for yourself this evening" (July 27, 1922, 1). Seeing and hearing, reading and listening, tuning in and being "there," the logic of newspaper-radio was manifest in the mobile radio tuner: "something like a combination of a steamboat on wheels and a prowling trench mortar. At the stern a coiled wire cylinder rears skyward. And on the roof, in front, perches a moveable horn."[11] Beginning late in July at Sunnyside, Toronto's newest amusement park, the Radio Car was dispatched nightly, drawing crowds of several hundred people away from the other amusements at hand. The remote concerts offered by the Radio Car—on occasion as far as Oakville, Barrie, and Oshawa, and in dozens of parks around the city—provided radio listening to the public who did not have receiving sets at home. It also provided a way to continue the habit of radio listening during long summer evenings when people were more likely to be outside than at home near the radio. Each afternoon, the *Star* printed that evening's radio program in capsule form, but the previous evening's radio broadcast was reviewed in eyewitness story form, specifically framed by the newsworthy arrival of the Radio Car to enliven the crowd at a park by transforming their leisure into listening. The Radio Car helped listener-readers to see radio through the spectacle of the technology in the form of a specially-made automobile.

Radio and Shifting Temporalities of Newspaper Reading

Newscasts were one of the final elements that the *Star* added to its daily CFCA radio program, late in August 1922. Commencing just in time for the "Radio Year" Exhibition of 1922, the addition of news and weather

---

11 A young Edward S. (Ted) Rogers, just twenty-two years old but already moving to the centre of the new industry, recalled operating the Radio Car as part of his work in 1922 for the Independent Telephone Company (Chaplin 2005, 72). This role is not mentioned at the time in the *Star* itself, however, although Rogers was profiled by the Radio Department for his pioneering amateur transmissions from suburban Newmarket, picked up in Scotland (September 20, 1922, 11).

transformed the CFCA "concerts" into a full-service program directly paralleling every section of the daily newspaper. By the end of the year, the newspaper-radio went one step further than the printed paper had ever ventured by broadcasting on Sundays (November 30, 1922, 17). Radio allowed temporal shifts in how the public engaged with news and conceived of the immediacy and informative capacity of the newspaper. The point is not that radio reports replaced newspaper stories, but that they supplemented print news by allowing occasional emergency broadcasts and updates on developing stories, as we will review shortly. The extension of the newspaper's news function to its radio service was hardly promoted as a momentous or innovative feature. Indeed, the first news bulletins were noted only after the fact as providing "an agreeable surprise to the listeners" (August 25, 1922, 7). Of course, the story of the first newscasts on CFCA appeared on the radio page (for the benefit of its radio fans) rather than the *Star*'s front page (for the benefit of its readers seeking news).

The broadcasting radius of several hundred miles now exceeded the geographic circulation of the newspaper itself. However, that very extension of the reach of the daily paper only drew attention to a temporal problem of newsreading that had been unimportant before 1922: a significant minority of the *Star*'s readership was outside the city and received their afternoon edition newspapers from Toronto by train late in the evening. Programming would soon start at noon and provide intermittent news bulletins throughout the day, a format first tested with eight-hour daily programming during the 1922 Exhibition. The benefit of broadcasting news bulletins was immediately evident; with no afternoon editions circulating on Labour Day, the *Star* Radio Car was described in the paper as "a very able substitute for the newspapers" (September 5, 1922, 8). *Star* broadcasts had always included closing stock market reports and evening reports of sports scores, not fearing a loss of sales from its own afternoon papers because these features scooped its morning edition competitors. Early morning papers, such as the Toronto *Globe*, had long arranged special express delivery trains throughout Southern Ontario to ensure regional readers received their papers before breakfast (begun in 1887; see for instance *Globe*, March 3, 1888). Now, in 1922, afternoon papers like the *Star* had a very different deadline: the radio news broadcast at 7 PM. It was simply not possible for the paper to be delivered to all readers before that time on a daily basis, certainly not in every direction for up to 300 miles or more. Just two weeks after the introduction of radio news, the *Star*'s Radio Department reported "This Town is Modern—Gets News by Radio," describing how the Erie Electric Company in Hagersville was copying the

CFCA live bulletins by shorthand in order to post them outside the shop each evening, "drawing big crowds" (September 22, 1922, 10). The article does not note or even contemplate how many fewer (or perhaps more) newspapers were sold by the local newsdealer.

Live on-location reporting became a CFCA specialty. In practice, this primarily meant sports events.[12] However, the epitome of breaking news and live broadcasting was election night reports of voting returns. In fact, election returns were some of the first broadcasts in the United States, notably for the first evening on Westinghouse's KDKA on November 2, 1920 (Douglas 1999, 166). Election nights had long been a carnival of sorts, sponsored by newspapers. Readers had grown accustomed to expecting something other than the immediate results in the newspaper the next day; with radio, the common festive gathering at the newspaper office could be made even more immediate, and extended across a vast region. Usually, the crowd gathered at the newspaper's head office was part of the story, just as CFCA's election night news was itself newsworthy beginning in 1922. The first election returns CFCA provided were for the British election of November 15, 1922, lasting until midnight, long after its regular program. For the Toronto municipal elections on New Year's Day 1923, the *Star* arranged for twenty-six "public bonfire points" all around the city, where fires warmed audiences gathered to hear the vote count through loudspeakers. The Radio Car was at Earlscourt Park on St. Clair, despite the winter night, and several movie theatres outside the city limits were reported to have installed loudspeakers to provide the radio returns to their audiences.

Later that year, a few days before an Ontario provincial election, the *Star*'s Radio Department asked all readers with loudspeakers to write, telephone, or wire the radio editor if they were planning a public election gathering for their neighbourhood (June 18, 1923, 8). The result was a network of forty receiving stations, over a quarter from outside the city, all listed in the paper as places to gather to hear the vote counts. Despite the growing convenience of staying home to hear election night returns on the radio, the *Star* head office downtown was still promoted as "one continuous attraction" consisting of "movies, music, and community singing." At the multimedia spectacle, election returns were animated with political cartoons and slide pictures, while the results themselves were "promptly

---

12 By February 1923, CFCA was broadcasting live commentary for hockey games by remote microphone, including the first such hockey play-by-play by Foster Hewitt, whose iconic line, "he scores," was featured in the *Star*'s promotion of CFCA's broadcast of the opening ceremonies and first game at Maple Leaf Gardens (February 9, 1922; November 12, 1931).

issued to the public through the medium of the presses, lantern slides, radio, and telephone" (June 23, 1923, 4). For the Canadian federal election of 1926, the *Star* "doubled its facilities" by co-operating with two King Street banks to provide a pair of stages where jazz orchestras and moving pictures offered entertainment as extra editions, while CFCA transmitted results of the vote (September 13, 1926, 1).

The idea of other media supplementing the printed page can be traced to the escalating spectacle of such intermediality on election nights, led by and featuring newspaper head offices—now extended well beyond downtown sites through radio broadcasting. Indeed, the prospect of such exceptional occasions was manifest in the perceived necessity to be ever ready for emergency broadcasting. The rarity of actual emergencies did not hinder the *Star* from profiling moments when its radio broadcasts kept the newspaper present in mind when print copies could not be delivered. A snowstorm in February 1924 was just such an event. CFCA effectively replaced routine news operations between the city and the rest of Ontario, and "at the request of the Canadian Press, broadcast a service for them to Ontario newspapers which were unable to receive dispatches in the ordinary way" (August 29, 1933, 6). Similar moments merited front page stories lauding how CFCA could overcome the limitations of normal newspaper reporting. Early in 1925, false rumours that an ill Prime Minister Mackenzie King had died could be followed closely in radio bulletins all weekend (February 23, 1925, 1). Just a week later, an unprecedented earthquake occurred near Toronto, and CFCA could broadcast reports immediately upon its impact, keeping the public informed and relieving its anxiety (March 2, 1925, 1). In both cases, the news broke Saturday evening, and CFCA could provide news throughout Sunday, more than a full day before Monday morning papers.

Since the first days of broadcasting, such instances of the utility and public service of radio bettering newsprint had elicited concern from editors and publishers. The spectre of being scooped by the immediacy of radio was central to a well-documented battle between radio and newspapers (Stamm 2011; Pratte 1993). Throughout the 1920s, bans were issued on radio using Canadian Press wire service, which was owned by a collective of newspaper publishers and had exclusive right to distribute Associated Press wire stories in Canada (Allen 2009). Such national policies were rarely upheld because Toronto newspapers in particular refused to abstain from bringing news to air on radio. No doubt this was instigated by the *Star*'s direct ownership of CFCA, but by the end of the decade the *Globe* was regularly supplying a newscast to Rogers' CFRB and the *Telegram* to Gooderham & Worts's CKGW (53). Initially in 1922, however, the *Star*'s competition on the air was less its rivals at newsstands in the city of Toronto than the freely available jazz and news of American airwaves. When it came to radio, competition in Toronto went beyond the city's own newspapers, as broadcasts from almost every point in the United States could be received in Southern Ontario better than any other point in Canada. While the *Star*'s radio page printed an unsurprising amount of publicity for American programming, the editors nonetheless maintained a surprising emphasis on CFCA's own reach south of the border, constantly calling for distant listeners to send in letters. "Canada's Finest Covers America" was a slogan offered by a listener-reader, and it became the station's slogan (December 16, 1922, 11). A whole chart of alternative slogans was submitted by another listener shortly afterward, appealing this time more directly to the interchangeable variety of the *Star* and its broadcasting, providing a slogan for each of Radio, News, Editorials, Sporting, Want Ads, and Advertising, and the result was: "Canada's Finest Circulation Attained" (February 10, 1923, 6).

Conclusion

CFCA was a short-lived radio station, as were most newspaper-radio stations in Canada. The station was shuttered in 1933 after only eleven years in operation. At the time, the *Star* noted that it had been an open supporter of the federal government's steps to nationalize radio broadcasting. Indeed, the *Star* continued to provide sponsored regional newscasts to the Toronto CBC affiliate until 1946, promoted daily on its own radio page. Although CFCA likely left little mark on the overall development of private radio broadcasting in Canada, the case is important for highlighting the changing ways publics were addressed in an increasingly diverse mediated landscape. CFCA's history is thus not specifically about the birth of radio; it problematizes mass media publics generally. The radio audience was already a newspaper readership, forged through the circulation of the paper on a daily basis to family homes in urban and suburban settings alike, across an entire metropolitan region. The temporal and spatial character of broadcasting was not without precedent. To the contrary, as we have pointed out, the character of radio listening at first complemented newspaper reading so closely as to make the radio page and radio program both supplements of the newspaper, engaged in organizing a newly-formed common public. By 1922, newspaper circulation already had temporal elements outside of the paper's periodicity, due especially to supplements like puzzles, cut-out toys, keepsake posters, coupons, and serial stories, all of which required saving the paper beyond its due date. The newspaper-radio station further supplemented the temporal rhythms of newsreading by adding an element of continuity between print editions, with evening hours and Sunday programs, the latter especially important in Canada (Gabriele 2011). Circulation itself is not merely an economic or empirical fact; it is a conceptual matter of the newspaper's social function. Although each form of media is often understood as having distinct audiences, the modern mediated public is always intermedial and understands each individuated medium through and in relation to the knowledge and leisure practices of the other.

Analytically, neither of the two forums for the *Star*'s publicity should be given priority as the primary site of the novelty of radio; the mass broadcast of the radio program was designed to be accompanied by the mass reading of the radio page, and vice versa. This helps further explain our opening question about the need for public concerts for those without home radio receivers in the first days of radio. Facilitating active listening did not simply expand the audience and market for the new medium and its consumer goods—as if radio suppliers' advertising were the main concern of the *Star*'s radio page. Taking newspaper reading beyond the page and transforming knowledge gleaned from reading into social activities was an organizing principle for the newspaper overall, especially for someone doubly charged with promotions and radio programming like the *Star*'s William Main Johnson. "Listening in" involved *writing in* and *reading up* the next day. Much like the act of reading the newspaper, the act of listening went well beyond the specific mediated moment. A sense of the entire mediated environment is necessary to capture the nuances of the shifting temporal and spatial organization of bodies, practices, and technologies. To write these histories requires an eye to how media worked together. Placed within an intermedial context, already both urban and suburban, the addition of radio to newspaper publishing makes sense.

References
Primary Sources

Johnson, William Main. Toronto Public Library. L33.

Secondary Sources

Allen, Gene. 2009. "Old Media, New Media, and Competition: Canadian Press and the Emergence of Radio News." In *Communicating in Canada's Past: Essays in Media History*, ed. Gene Allen and Daniel J. Robinson, 47–77. Toronto: University of Toronto Press.
Altman, Rick. 2004. *Silent Film Sound*. New York: Columbia University Press.
Anderson, Benedict. 1983. *Imagined Communities: Reflections on the Origins and Spread of Nationalism*. New York: Verso.

Barnhurst, Kevin, and John Nerone. 2001. *The Form of News: A History*. New York: Guildford Press.

Chaplin, Maurice. 2005. "Just Plug In—Then Tune In: The First Commercial Light-socket Operated Radio Receivers from Rogers Radio Ltd." In *The Early Development of Radio in Canada, 1901–1930*, ed. Robert P. Murray, 69–92. Chandler, AZ: Sonoran Publishing.

Dooley, Patricia. 2007. *The Technology of Journalism: Cultural Agents, Cultural Icons*. Evanston, IL: Northwestern University Press.

Dorland, Michael. 1996. "Introduction." In *The Cultural Industries in Canada: Problems, Policies and Prospects*, ed. Michael Dorland, ix–xiii. Toronto: James Lorimer & Co. Publishers.

Douglas, Susan. 1987. *Inventing American Broadcasting, 1899–1922*. Baltimore: Johns Hopkins University Press.

———. 1999. *Listening In: Radio and the American Imagination*. New York: Random House.

Gabriele, Sandra. 2011. "Cross-border Transgressions: The American Sunday Newspaper, the Lord's Day Alliance and the Reading Public, 1890–1916." *Topia: A Journal of Cultural Studies* 25 (2): 115–132.

Gabriele, Sandra, and Paul S. Moore. 2009. "*The Globe* on Saturday, *The World* on Sunday: Toronto Weekend Editions and the Influence of the American Sunday Paper, 1886–1895." *Canadian Journal of Communication* 34 (3): 337–358.

Gitelman, Lisa. 2006. *Always Already New: Media, History and the Data of Culture*. Cambridge, MA: MIT Press.

Gordon, Ian. 1998. *Comic Strips and Consumer Culture, 1890–1945*. Washington, DC: Smithsonian Institute.

Hilmes, Michele. 1997. *Radio Voices: American Broadcasting, 1922–1952*. Minneapolis: University of Minnesota Press.

Jackson, John. 1999. "Canadian Radio Research: An Introduction." *Journal of Radio Studies* 6 (1): 116–120.

Kuffert, Len. 2009. "'What Do You Expect of This Friend?' Canadian Radio and the Intimacy of Broadcasting." *Media History* 15 (3): 303–319.

MacLennan, Anne. 2005. "American Network Broadcasting, the CBC, and Canadian Radio Stations during the 1930s: A Content Analysis." *Journal of Radio Studies* 12 (1): 85–103.

———. 2008. "Women, Radio and the Depression: A 'Captive' Audience from Household Hints to Story Time and Serials." *Women's Studies* 37 (6): 616–633.

———. 2012. "Reading Radio: The Intersection between Radio and Newspaper for the Canadian Radio Listener in the 1930s." In *Radio and Society: New Thinking for an Old Media*, ed. Matthew Mollgaard, 16–29. Newcastle-upon-Tyne: Cambridge Scholars Publishing.

Martin, Michèle. 2006. *Images at War: Illustrated Periodicals and Constructed Nations*. Toronto: University of Toronto Press.

Marvin, Carolyn. 1988. *When Old technologies Were New: Thinking about Electric Communication in the Late Nineteenth Century*. New York: Oxford University Press.

Méchoulan, Éric. 2003. "Intermédialités: le temps des illusions perdues." *Intermédialités / Intermediality* 1: 9–27.

Nord, David Paul. 2001. *Communities of Journalism: A History of American Newspapers and tTheir Readers*. Urbana: University of Illinois Press.

Pratte, Alf. 1993. "Going Along for the Ride on the Prosperity Bandwagon: Peaceful Annexation Not War between the Editors and Radio, 1923–1941." *Journal of Radio Studies* 2 (3): 123–139.

Raboy, Marc. 1990. *Missed Opportunities: The Story of Canada's Broadcasting Policy*. Montreal: McGill-Queen's University Press.

Razlogova, Elena. 2011. *The Listener's Voice: Early Radio and the American Public*. Philadelphia: University of Pennsylvania Press.

Rutherford, Paul. 1982. *A Victorian Authority: The Daily Press in Late Nineteenth-century Canada*. Toronto: University of Toronto Press.

Skinner, David. 2005. "Divided Loyalties: The Early Development of Canada's 'Single' Broadcasting System." *Journal of Radio Studies* 12 (1): 136–155.

Stamm, Michael. 2011. *Sound Business: Newspapers, Radio and the Politics of New Media*. Philadelphia: University of Pennsylvania Press.

Vipond, Mary. 1992. *Listening In: The First Decade of Canadian Broadcasting, 1922–1932*. Montreal: McGill-Queen's University Press.

———. 1999. "The Continental Marketplace: Authority, Advertisers and Audiences in Canadian News Broadcasting, 1932–1936." *Journal of Radio Studies* 6 (1): 169–184.

———. 2009. "Whence and Whither: The Historiography of Canadian Broadcasting." In *Communicating in Canada's Past: Essays in Media History*, ed. Gene Allen and Daniel J. Robinson, 233–256. Toronto: University of Toronto Press.

———. 2010. "The Royal Tour of 1939 as Media Event." *Canadian Journal of Communication* 35 (1): 149–172.

Wagman, Ira. 2010. "On the Policy Reflex in Canadian Communication Studies." *Canadian Journal of Communication* 35 (4): 619–630.

Walden, Keith. 1997. *Becoming Modern in Toronto: The Industrial Exhibition and the Shaping of Late Victorian Culture*. Toronto: University of Toronto Press.

Warner, Michael. 2002. "Publics and Counterpublics." *Public Culture* 14 (1): 49–90.

Webb, Jeff. 2008. *The Voice of Newfoundland: A Social History of the Broadcasting Corporation of Newfoundland, 1939–1949*. Toronto: University of Toronto Press.

List of Contributors

Jeff Boggs (PhD Geography, UCLA 2005) is an Associate Professor at Brock University in Geography and the MA Program in Popular Culture. His research interests centre on the locational dynamics of cultural industries, with a focus on the book trade.

Olivier Côté holds a PhD in Canadian history from Laval University. He specializes in the study of media representations of identity. His thesis, to be published by Les éditions du Septentrion in winter 2013, focuses on the context of production, dissemination, and reception of media and the popular series *Canada: A People's History*. He has also published an article in the the prestigious French magazine *Hermès* on Quebecois and Canadian media coverage of French and Dutch referenda on the European integration project.

Greig de Peuter is an Assistant Professor in the Department of Communication Studies at Wilfrid Laurier University. He is the co-author, with Nick Dyer-Witheford, of *Games of Empire: Global Capitalism and Video Games*.

Christopher Dornan has taught at Carleton University since 1987, where he has served as Director of the School of Journalism and Communication (1997–2006) and Director of the Arthur Kroeger College of Public Affairs (2007–13). He is also co-editor (with Jon Pammett) of *The Canadian Federal Election of 2011* and four previous volumes in this series. A collection of his essays and criticism can be found at www.educatedguesses.ca.

Zoë Druick is an Associate Professor in the School of Communication at Simon Fraser University. Her books include *Projecting Canada:*

Documentary Film and Government Policy at the National Film Board (2007), *Programming Reality: Perspectives on English-Canadian Television* (2008), and *Allan King's A Married Couple* (2010). She has also published numerous articles on reality-based and educational media in journals such as *Screen, Television and New Media, Canadian Journal of Communication, Canadian Journal of Film Studies,* and *Studies in Documentary,* as well as a number of anthologies.

Sandra Gabriele is Assistant Professor of Communication Studies at Concordia University. A forthcoming book with Paul Moore, *The Sunday Paper,* explores the intermedial circulation of illustrated weekend newspapers in North America. In addition to co-editing *Intersections of Media and Communications* (2011) with Will Straw and Ira Wagman, her histories of journalism in Canada have appeared in *Topia, Aether,* and *Journalism: Theory, Practice & Criticism.* New research traces the democratic and epistemological potential lost and gained by the rise of historical newspaper databases, and thus the relationship between access, policy, and preservation practices such as microfilming.

Mark Hayward teaches in the Department of Communication Studies at Wilfrid Laurier University. He has published essays on diaspora, popular culture, and critical theory.

Paul S. Moore is Associate Professor of Communication and Culture at Ryerson University. Research with Sandra Gabriele, including a forthcoming book, *The Sunday Paper,* explores the intermedial circulation of illustrated weekend newspapers in North America. His histories of film exhibition in North America have appeared in the *Canadian Journal of Film Studies,* as chapters in *Explorations in New Cinema History* and *A Companion to Early Cinema,* and the book *Now Playing* (2008), winner of the Robinson Prize of the Canadian Communication Association.

Daniel J. Paré is an Associate Professor with the Department of Communication at the University of Ottawa, and is cross-appointed into both the School of Information Studies and the Institute for Science, Society and Policy. His primary areas of research and policy-related work focus on social, economic, political, and technological issues arising from innovations in information and communication technologies.

Jeremy Shtern is Assistant Professor in the Department of Communication, University of Ottawa. His research focuses on the conflux of communication policy, globalization, digital technologies, and, increasingly, creative labour in the cultural industries. Amongst various research contributions, Jeremy has co-authored two recent books, *Media Divides: Communication Rights and the Right to Communicate in Canada* (2010, UBC Press, with Marc Raboy) and *Digital Solidarities, Communication Policy and Multi-stakeholder Global Governance* (2010, Peter Lang, with Marc Raboy and Normand Landry).

Richard Sutherland is assistant professor in the Department of Policy Studies at Mount Royal University in Calgary, Alberta. Prior to academic life he worked in the Canadian music industry in a number of roles, from journalist to label manager to policy researcher. His research interests are the music industry and cultural policy. In particular, his work has focused on the Canadian music industry and government policy from the 1960s onward.

Dwayne Winseck is Professor at the School of Journalism and Communication at Carleton University. His work focuses on media history, media ownership and concentration, surveillance, and democracy. He is a regular columnist for the *Globe and Mail* and maintains a well-regarded blog, *Mediamorphis* (dwmw.wordpress.com). His last book (co-authored with Robert Pike), *Communication and Empire: Media, Markets and Globalization, 1860–1930* (Duke University, 2007), won the Canadian Communication Association's book of the year prize in 2008. He is also the co-editor of a new volume, with Dal Yong Jin: *Political Economies of the Media: the Transformation of the Global Media Industries* (2011).

Index